Out of Nantucket

Out of Nantucket
Two accounts of whalers and whaling

A Narrative of the Mutiny on
Board the Ship Globe
William Lay and Cyrus M. Hussey

Memorandum
William Rotch

Out of Nantucket
Two accounts of whalers and whaling
*A Narrative of the Mutiny on
Board the Ship Globe*
by William Lay and Cyrus M. Hussey
and
Memorandum
by William Rotch

First published under the titles
*A Narrative of the Mutiny on
Board the Ship Globes*
and
Memorandum

Leonaur is an imprint of Oakpast Ltd

Copyright in this form © 2010 Oakpast Ltd

ISBN: 978-0-85706-080-8 (hardcover)
ISBN: 978-0-85706-079-2 (softcover)

http://www.leonaur.com

Publisher's Notes

In the interests of authenticity, the spellings, grammar and place names used have been retained from the original editions.

The opinions of the authors represent a view of events in which he was a participant related from his own perspective, as such the text is relevant as an historical document.

The views expressed in this book are not necessarily those of the publisher.

Contents

A Narrative of the Mutiny on Board the Ship Globe   7

Memorandum   87

A Narrative of the Mutiny on Board the Ship Globe

William Lay and Cyrus M. Hussey

Contents

Introduction	15
Commencement of the Mutiny	19
Take Command of the Ship	26
Surviving the Massacre	36
After the Massacre	45
Adventures of William Lay	48
Escape From the Island	57
Observations of the Islands Visited	70
Cyrus Hussey's Journal	74
Journal Continued	81

To John Percival, Esq. of the U. S. Navy,
Who, under the auspices of Government,
visited the Mulgrave islands,
to release the survivors
of the Ship *Globe*'s crew,
and extended to them every attention
their unhappy situation required—
the following Narrative is
most respectfully dedicated, by

William Lay, &
Cyrus M. Hussey,
The Authors.

The Young Mutineer

*His sun rose unclouded and brightly it shone
In the pride of the morning and promised a noon
Of glory and gladness; it sank to the Hood
In blackness and blindness, and tarnished by blood.
Disowned and dishonoured its last gloomy glare
Was shed on the grave of the young mutineer.*

*Tho' beardless his cheek, yet his was a soul
That knew not a master, that brooked no control;
Tho' beardless his cheeks, yet his was a hand
Acquainted with daggers; a voice to command.
An eye that wept, a heart without fear.
Were the pride and the boast of the young mutineer.*

*He lies on the beach of a lone desert isle.
His dirge the green billows are chanting the while,
As they in wild tumult, roll over his head.
And wash the high rock, that marks out his wet bed;
There lies with a heart that ne'er knew a fear.
The mangled remains of the young mutineer.*

*He lies on the beach, the cold waters beside.
And dreadful and dark was the death that he died
No mother mourns o'er him, no fond fair one weeps
Where far from the land of his father he sleeps.
But the rude swelling wave, and the sea birds career
On the wet sandy grave of the young mutineer.*

*He lies on the beach by a comrade in guilt.
His forehead was cloven, his best blood spilt.
The cries of his victims have risen to God,
And their wailings were quenched
In the murderer's blood.
He fell without mourners, for none dropped a tear
O'er the mangled remains of the young mutineer.*

He lies on the beach, where the weeds and the shells,
Mark the bounds of the sea, where in tumult it swells
They scooped him a grave, and there laid him at rest.
And heaped the wet sand on his bare, bloody breast.
And they rolled a huge rock and planted it there
To mark the lone grave of the young mutineer.

In years that are coming the seamen will tell
Of murders and murdered, and murderer's yells.
The tale, the lone watch of night will beguile
When they sail by the shores of that desolate isle.
And their beacon shall be, as they thitherward steer.
The black rock on the grave of the young mutineer.

Henry Glover.

Introduction

Formerly whales were principally taken in the North Seas: the largest were generally found about Spitzbergen, or Greenland, some of them measuring ninety feet in length. At the commencement of the hazardous enterprise of killing whales, before they had been disturbed by man, they were so numerous in the bays and harbours, that when taken the blubber was for the most part boiled into oil upon the contiguous coast.

The pure oil and whale bone were only preserved in those days; consequently a ship could carry home the product of a greater number of whales than a ship of the same size now can. Indeed, so plentiful were the whales in those seas, and taken with such facility, that the ships employed, were not sufficient to carry home the oil and bone, and other ships were often sent to bring home the surplus quantity. But the coasts of these countries, were soon visited by ships from Denmark, Hamburgh, and Holland, as well as from England; and from frequently being killed in the shoal water near the coasts, the whales gradually receded from the shores, and have since been found only in deeper water, and at a much greater distance from the land.

In the earlier stages of the whale fishery, of which we are now treating, the ships were generally on the whaling waters, early in May, and whether successful or not, they were obliged to commence their return by the succeeding August, to avoid the early accumulation of ice in those seas. But it not unfrequently happened, that ships procured and returned with a cargo in the months of June and July, making a voyage only about three months, whereas, a voyage to the Pacific Ocean is now often protracted to three years!

Among the early whalers it was customary to have six boats to a ship, and six men to a boat, besides the harpooner. What at that time was considered an improved method in killing whales, consisted in discharging the harpoon, from a kind of swivel; but it was soon found to be attended with too much inconvenience to be much practised, and the muscular arms and steady nerves of the harpooner, have ever since performed the daring duty, of first striking the whale. The ropes attached to the harpoon, used to be about 200 fathoms in length, and some instances occurred, that all the lines belonging to six boats, were fastened together and ran out by one whale, the animal descending in nearly a perpendicular line from the surface. Instead of going prepared to bring home a ship load of oil, it was customary to bring only the blubber, and instead of trying the oil out and putting it into casks on board, the fat of the whale was cut up into suitable pieces, pressed hard in tubs carried out for the purpose, and in this situation was the return cargo received at home.

Of so great consequence was the whale fishery considered to Great Britain, that a bounty of 40s. for every ton, when the ship was 200 tons, or upwards, was given to the crews of ships engaged in that business in the Greenland seas, under certain conditions. But this bounty was found to draw too largely upon the treasury; and while the subject was under discussion in the British Parliament, in 1786, it was stated that the sums which that country had paid in bounties to the Greenland fishers, amounted to 1,265,461 pounds sterling. Six thousand seamen were employed in that fishery, and each cost the government £13 10s. *per annum.* The great encouragement given to that branch of commerce, caused so large a number to engage in it, that the oil market became glutted, and it was found necessary to export considerable quantities.

In 1786, the number of British ships engaged in the whale fishery to Davis's Strait and the Greenland seas, was 139, besides 15 from Scotland. In 1787, notwithstanding the bounty had been diminished, the number of English ships was 217, and the following year 222.

The charter right of the island of Nantucket, was bought by Thomas Mayhew, of Watertown, of Joseph Ferrick, steward to Lord Sterling, in 1641; and afterwards sold to Tristram Coffin, and his associates, who settled upon it in 1659. On the 10th of May, 1660, Sachems, Wonnook, and Nickannoose, for and in behalf of the nations of the island, in consideration of the sum of £25, conveyed by deed, about half of the island, to the first ten purchasers, who afterwards took in other associates.

Whaling from Nantucket, was first carried on from the shore in boats. In 1672, James Loper entered into a contract with the inhabitants of the island, for the purpose of prosecuting the whale fishery, by which it appears that James Loper agreed to be one third in the enterprise, and sundry other people of the island, the other two thirds, in everything connected with the undertaking. It was further stipulated, that for every whale killed by any one of the contracting party, the town should receive five shillings, and for the encouragement of James Loper, the town granted him ten acres of land in some convenient situation, and liberty for the commonage of three cows, twenty sheep and one horse, with necessary wood and water for his use, on condition that he should follow the trade of whaling for two years, build upon his land, &c. &c.

Thus it will be seen that the commencement of whaling at Nantucket, was on a very small scale, and practised only along the shores of the island;—whereas, at this time, our ships leave no seas unexplored in pursuit of these monsters of the deep. We might pursue the subject through the various stages of improvement up to this time, but it would swell this introduction beyond the limits designed. It is proper, however, to observe that the present number of ships employed in the whale fishery from Nantucket, is about 70, averaging about 350 tons each, and manned by about 1500 seamen.

Chapter 1

Commencement of the Mutiny

The Ship *Globe*, on board of which vessel occurred the horrid transactions we are about to relate, belonged to the island of Nantucket; she was owned by Messrs. C. Mitchell, & Co. and other merchants of that place; and commanded on this voyage by Thomas Worth, of Edgartown, Martha's Vineyard. William Beetle, (mate,) John Lumbert, (2nd mate,) Nathaniel Fisher, (3rd mate,) Gilbert Smith, (boat steerer,) Samuel B. Comstock, (boat steerer,). Stephen Kidder, (seaman,) Peter C. Kidder, (seaman,). Columbus Worth, (seaman,) Rowland Jones, (seaman,) John Cleveland, (seaman,). Constant Lewis, (seaman,) Holden Henman, (seaman,) Jeremiah Ingham, (seaman,) Joseph Ignasius Prass, (seaman,). Cyrus M. Hussey, (cooper,) Rowland Coffin, (cooper,) George Comstock, (seaman,) and William Lay, (seaman.)

On the 15th day of December, we sailed from Edgarton, on a whaling voyage, to the Pacific Ocean, but in working out, having carried away the cross-jack-yard, we returned to port, and after having refitted and sent aloft another, we sailed again on the 19th, and on the same day anchored in Holmes' Hole. On the following day a favourable opportunity offering to proceed to sea, we got under way, and after having cleared the land, discharged the pilot, made sail, and performed the necessary duties of stowing the anchors, unbending and coiling away the cables, &c.

On the 1st of January 1823, we experienced a heavy gale from N. W. which was but the first in the catalogue of difficulties we were fated to encounter.—As this was our first trial of a seaman's life, the scene presented to our view, "mid the howling storm," was one of terrific grandeur, as well as of real danger. But as the ship scudded well, and the wind was fair, she was kept before it, under a close reefed main-top-sail and fore-sail, although during the gale, which lasted forty-eight hours, the sea frequently threatened to board us, which was prevented by the skilful management of the helm.

On the 9th of January we made the Cape Verd islands, bearing S. W. twenty-five miles distant, and on the 17th, crossed the Equator. On the 29th of the same month we saw sperm whales, lowered our boats, and succeeded in taking one; the blubber of which, when boiled out, yielded us seventy-five barrels of oil. Pursuing our voyage, on the twenty-third of February we passed the Falkland islands, and about the 5th of March, doubled the great promontory of South America, Cape Horn, and stood to the Northward.

We saw whales once only before we reached the Sandwich islands, which we made on the first of May early in the morning. When drawing in with the island of Hawaii about four in the afternoon, the man at the mast head gave notice that he saw a shoal of black fish on the lee bow; which we soon found to be canoes on their way to meet us. It falling calm at this time prevented their getting along side until nightfall, which they did, at a distance of more than three leagues from the land.

We received from them a very welcome supply of potatoes, sugar cane, yams, cocoanuts, bananas, fish, &c. for which we gave them in return, pieces of iron hoop, nails, and similar articles. We stood off and on during the next day, and after obtaining a sufficient supply of vegetables and fruit, we shaped our course for Oahu, at which place we arrived on the following day, and after lying there twenty hours, sailed for the coast of Japan, in company with the whaling ships *Palladium* of Boston, and *Pocahontas* of Falmouth; from which ships we parted company when two days out.—After cruising in the Japan seas several months, and obtaining five hundred and fifty barrels of oil, we again shaped our course for the Sandwich islands, to obtain a supply of vegetables, &c.

While lying at Oahu, six of the men deserted in the night; two of them having been retaken were put in irons, but one of them having found means to divest himself of his irons, set the other at liberty, and both escaped.

To supply their places, we shipped the following persons, *viz*: Silas Payne, John Oliver, Anthony Hanson, a native of Oahu, Wm. Humphries, a black man, and steward, and Thomas Lilliston.—Having accommodated ourselves with as many vegetables and much fruit as could be preserved, we again put to sea, fondly anticipating a successful cruise, and a speedy and happy meeting with our friends.

After leaving Oahu we ran to the south of the Equator, and after cruising a short time for whales without much success, we steered for Fannings island, which lies in lat. 3°, 49' N. and long. 158°, 29' W. While cruising off this island an event occurred which, whether we consider the want of motives, or the cold blooded and obstinate cruelty with which it was perpetrated, has not often been equalled.—We speak of the want of motives, because, although some occurrences which we shall mention, had given the crew some ground for dissatisfaction, there had been no abuse or severity which could in the least degree excuse or palliate so barbarous a mode of redress and revenge.

During our cruise to Japan the season before, many complaints were uttered by the crew among themselves, with respect to the manner and quantity in which they received their meat, the quantity sometimes being more than sufficient for the number of men, and at others not enough to supply the ship's company; and it is fair to presume, that the most dissatisfied, deserted the ship at Oahu.

But the reader will no doubt consider it superfluous for us to attempt an unrequired vindication of the conduct of the officers of the *Globe* whose aim was to maintain a correct discipline, which should result in the furtherance of the voyage and be a benefit to all concerned, more especially when he is informed, that part of the men shipped at Oahu, in the room of the deserters, were abandoned wretches, who frequently were the cause of severe reprimands from the officers, and in one instance one of them received a severe flogging.

The reader will also please to bear in mind, that Samuel B. Comstock, the ringleader of the mutiny, was an officer, (being a boat-steerer,) and as is customary, ate in the cabin. The conduct and deportment of the captain towards this individual, was always decorous and gentlemanly, a proof of intentions long premeditated to destroy the ship. Some of the crew were determined to leave the ship provided she touched at Fannings island, and we believe had concerted a plan of escape, but of which the perpetration of a deed chilling to humanity, precluded the necessity.

We were at this time in company with the ship *Lyra*, of New-Bedford, the captain of which, had been on board the *Globe* during the most of the day, but had returned in the evening to his own ship. An agreement had been made by him with the captain of the *Globe*, to set a light at midnight as a signal for tacking. It may not be amiss to acquaint the reader of the manner in which whalemen keep watch during the night. They generally carry three boats, though some carry four, five, and sometimes six, the *Globe*, however, being of the class carrying three. The captain, mate, and second mate stand no watch except there is blubber to be boiled; the boat-steerers taking charge of the watch and managing the ship with their respective boats crews, and in this instance dividing the night into three parts, each taking a third.

It so happened that Smith after keeping the first watch, was relieved by Comstock, (whom we shall call by his surname in contradistinction to his brother George) and the waist boat's crew, and the former watch retired below to their births and hammocks. George Comstock took the helm, and during his trick, received orders from his brother to "keep the ship a good full," swearing that the ship was too nigh the wind. When his time at the helm had expired he took the rattle, (an instrument used by whalemen, to announce the expiration of the hour, the watch, &c.) and began to shake it, when Comstock came to him, and in the most peremptory manner, ordered him to desist, saying "if you make the least damn bit of noise I'll send you to hell!"

He then lighted a lamp and went into the steerage. George becoming alarmed at this conduct of his unnatural brother, again took the rattle for the purpose of alarming someone; Comstock arrived in time to prevent him, and with threatenings dark and diabolical, so congealed the blood of his trembling brother, that even had he possessed the power of alarming the unconscious and fated victims below, his life would have been the forfeit of his temerity!

Comstock, now laid something heavy upon a small work bench near the cabin gangway, which was afterwards found to be a boarding knife. It is an instrument used by whalers to cut the blubber when hoisting it in, is about four feet in length, two or three inches wide, and necessarily kept very sharp, and for greater convenience when in use, is two edged.

In giving a detail of this chilling transaction, we shall be guided by the description given of it by the younger Comstock, who, as has been observed, was upon deck at the time, and afterwards learned several particulars from his brother, to whom alone they could have been known. Comstock went down into the cabin, accompanied by Silas Payne or Paine, of Sag-Harbour, John Oliver, of Shields, Eng., William Humphries, (the steward) of Philadelphia, and Thomas Lilliston; the latter, however, went no farther than the cabin gangway, and then ran forward and turned in.

According to his own story he did not think they would attempt to put their designs in execution, until he saw them actually descending into the cabin, having gone so far, to use his own expression, to show himself as brave as any of them. But we believe he had not the smallest idea of assisting the villains. Comstock entered the cabin so silently as not to be perceived by the man at the helm, who was first apprised of his having begun the work of death, by the sound of a heavy blow with an axe, which he distinctly heard.

The captain was asleep in a hammock, suspended in the cabin, his state room being uncomfortably warm; Comstock approaching him with the axe, struck him a blow upon the head, which was nearly severed in two by the first stroke! After repeating the blow, he ran to Payne, who it seems was stationed with the before mentioned boarding knife, to attack the mate, as soon as the captain was killed. At this instant, Payne making a thrust at the mate, he awoke, and terrified, exclaimed, "what! what! what! Is this——Oh! Payne! Oh! Comstock! Don't kill me, don't; have I not always——"

Here Comstock interrupted him, saying, "Yes! you have always been a d——d rascal; you tell lies of me out of the ship will you? It's a d——d good time to beg now, but you're too late," here the mate sprang, and grasped him by the throat. In the scuffle, the light which Comstock held in his hand was knocked out, and the axe fell from his hand; but the grasp of Mr. Beetle upon his throat, did not prevent him from making Payne understand that his weapon was lost, who felt about until he found it, and having given it to Comstock, he managed to strike him a blow upon the head, which fractured his skull; when he fell into the pantry where he lay groaning until despatched by Comstock! The steward held a light at this time, while Oliver put in a blow as often as possible!

The second and third mates, fastened in their state rooms, lay in their births listening, fearing to speak, and being ignorant of the numerical strength of the mutineers, and unarmed, thought it best to wait the dreadful issue, hoping that their lives might yet be spared.

Comstock leaving a watch at the second mate's door, went upon deck to light another lamp at the binnacle, it having been again accidentally extinguished. He was there asked by his terrified brother, whose agony of mind we will not attempt to portray, if he intended to hurt Smith, the other boat-steerer. He replied that he did; and inquired where he was. George fearing that Smith would be immediately pursued, said he had not seen him.—Comstock then perceiving his brother to be shedding tears, asked sternly, "What are you crying about?"

"I am afraid," replied George, "that they will hurt me!"

"I will hurt you," said he, "if you talk in that manner!"

But the work of death was not yet finished. Comstock, took his light into the cabin, and made preparations for attacking the second and third mates, Mr. Fisher, and Mr. Lumbert. After loading two muskets, he fired one through the door, in the direction as near as he could judge of the officers, and then inquired if either was shot! Fisher replied, "yes, I am shot in the mouth!" Previous to his shooting Fisher, Lumbert asked if he was going to kill him? To which he answered with apparent unconcern, "Oh no, I guess not."

They now opened the door, and Comstock making a pass at Mr. Lumbert, missed him, and fell into the state room. Mr. Lumbert collared him, but he escaped from his hands. Mr. Fisher had got the gun, and actually presented the bayonet to the monster's heart! But Comstock assuring him that his life should be spared if he gave it up, he did so; when Comstock immediately ran Mr. Lumbert through the body several times!!

He then turned to Mr. Fisher, and told him there was no hope for him!!—"You have got to die," said he, "remember the scrape you got me into, when in company with the *Enterprise* of Nantucket." The "scrape" alluded to, was as follows. Comstock came up to Mr. Fisher to wrestle with him.—Fisher being the most athletic of the two, handled him with so much ease, that Comstock in a fit of passion struck him. At this Fisher seized him, and laid him upon deck several times in a pretty rough manner.

Comstock then made some violent threats, which Fisher paid no attention to, but which now fell upon his soul with all the horrors of reality. Finding his cruel enemy deaf to his remonstrances, and entreaties, he said, "If there is no hope, I will at least die like a man!" and having by order of Comstock, turned back to, said in a firm voice, "I am ready!!"

Comstock then put the muzzle of the gun to his head, and fired, which instantly put an end to his existence!—Mr. Lumbert, during this time, was begging for life, although no doubt mortally wounded. Comstock, turned to him and said, "I am a bloody man! I have a bloody hand and will be avenged!" and again ran him through the body with a bayonet!

He then begged for a little water; "I'll give you water," said he, and once more plunging the weapon in his body, left him for dead!

Thus it appears that this more than demon, murdered with his own hand, the whole! Gladly would we wash from "memory's waste" all remembrance of that bloody night. The compassionate reader, however, whose heart sickens within him, at the perusal, as does ours at the recital, of this tale of woe, will not, we hope, disapprove our publishing these melancholy facts to the world. As, through the boundless mercy of Providence, we have been restored, to the bosom of our families and homes, we deemed it a duty we owe to the world, to record our "unvarnished tale."

Chapter 2

Take Command of the Ship

Smith, the other boat-steerer, who had been marked as one of the victims, on hearing the noise in the cabin, went aft, apprehending an altercation between the captain and some of the other officers, little dreaming that innocent blood was flowing in torrents. But what was his astonishment, when he beheld Comstock, brandishing the boarding knife, and heard him exclaim, "I am the bloody man, and will have revenge!" Horror struck, he hurried forward, and asked the crew in the forecastle, what he should do. Some urged him to secrete himself in the hold, others to go aloft until Comstock's rage should be abated; but alas! the reflection that the ship afforded no secure hiding place, determined him to confront the ringleader, and if he could not save his life by fair means, to sell it dearly!

He was soon called for by Comstock, who upon meeting him, threw his bloody arms around his neck, and embracing him, said, "you are going to be with us, are you not?"

The reader will discover the good policy of Smith when he unhesitatingly answered, "Oh, yes, I will do anything you require."

All hands were now called to make sail, and a light at the same time was set as a signal for the *Lyra* to tack;—while the *Globe* was kept upon the same tack, which very soon caused a separation of the two ships. All the reefs were turned out, top-gallant-sails set, and all sail made on the ship, the wind being quite light.

The mutineers then threw the body of the captain overboard, after wantonly piercing his bowels with a boarding knife, which was driven with an axe, until the point protruded from his throat!! In Mr. Beetle, the mate, the lamp of life had not entirely gone out, but he was committed to the deep.

Orders were next given to have the bodies of Mr. Fisher, and Mr.

Lumbert brought up. A rope was fastened to Fisher's neck, by which he was hauled upon deck. A rope was made fast to Mr. Lumbert's feet, and in this way was he got upon deck, but when in the act of being thrown from the ship, he caught the plank-shear; and appealed to Comstock, reminding him of his promise to save him, but in vain; for the monster forced him from his hold, and he fell into the sea! As he appeared to be yet capable of swimming, a boat was ordered to be lowered, to pursue and finish him, fearing he might be picked up by the *Lyra*; which order was as soon countermanded as given, fearing, no doubt, a desertion of his murderous companions.

We will now present the reader, with a journal of our passage to the Mulgrave Islands, for which group we shaped our course.

1824, *Jan. 26th*. At 2 a. m. from being nearly calm a light breeze sprung up, which increased to a fresh breeze by 4 a. m. This day cleaned out the cabin, which was a scene of blood and destruction of which the recollection at this day chills the blood in our veins.—Everything bearing marks of the murder, was brought on deck and washed.

Lat. 5° 50' N. Long. 159° 13' W.

Jan. 27th. These twenty-four hours commenced with moderate breezes from the eastward. Middle and latter part calm. Employed in cleaning the small arms which were fifteen in number, and making cartridge boxes.

Lat. 30° 45' N. Long. 160° 45' W.

Jan. 28th. This day experienced fine weather, and light breezes from N. by W. The black steward was hung for the following crime.

George Comstock who was appointed steward after the mutiny, and business calling him into the cabin, he saw the former steward, now called the purser, engaged in loading a pistol. He asked him what he was doing that for. His reply was, "I have heard something very strange, and I'm going to be ready for it." This information was immediately carried to Comstock, who called to Payne, now mate, and bid him follow him.

On entering the cabin they saw Humphreys, still standing with the pistol in his hand. On being demanded what he was going to do with it, he said he had heard something which made him afraid of his life!

Comstock told him if he had heard anything, that he ought to have come to him, and let him know, before he began loading pistols. He then demanded to know, what he had heard. Humphreys answered at first in a very suspicious and ambiguous manner, but at length said, that Gilbert Smith, the boat-steerer who was saved, and Peter Kidder, were going to retake the ship. This appeared highly improbable, but they were summoned to attend a council at which Comstock presided, and asked if they had entertained any such intentions. They positively denied ever having had conversation upon the subject.

All this took place in the evening. The next morning the parties were summoned, and a jury of two men called. Humphreys under a guard of six men, armed with muskets, was arraigned, and Smith and Kidder, seated upon a chest near him. The prisoner was asked a few questions touching his intentions, which he answered but low and indistinctly. The trial, if it may be so called, had progressed thus far, when Comstock made a speech in the following words.

> It appears that William Humphreys has been accused guilty, of a treacherous and base act, in loading a pistol for the purpose of shooting Mr. Payne and myself. Having been tried the jury will now give in their verdict, whether Guilty or Not Guilty. If guilty he shall be hanged to a studding-sail boom, rigged out eight feet upon the fore-yard, but if found not guilty, Smith and Kidder, shall be hung upon the aforementioned gallows!

But the doom of Humphreys had been sealed the night before, and kept secret except from the jury, who returned a verdict of guilty.— Preparations were immediately made for his execution! His watch was taken from him, and he was then taken forward and seated upon the rail, with a cap drawn over his face, and the rope placed round his neck.

Every man was ordered to take hold of the execution rope, to be ready to run him up when Comstock should give the signal, by ringing the ship's bell!

He was now asked if he had anything to say, as he had but fourteen seconds to live! He began by saying, "little did I think I was born to come to this———;" the bell struck! and he was immediately swung to the yard-arm! He died without a struggle; and after he had hung a few minutes, the rope was cut, to let him fall overboard, but getting entangled aloft, the body was towed some distance alongside, when a runner hook,[1] was attached to it, to sink it, when the rope was again cut and the body disappeared. His chest was now overhauled, and sixteen dollars in *specie* found, which he had taken from the captain's trunk. Thus ended the life of one of the mutineers, while the blood

---

1. A large hook used when hoisting in the blubber.

of innocent victims was scarcely washed from his hands, much less the guilty stain from his soul.

Feb. 7th. These twenty-four hours commenced with thick squally weather. Middle part clear and fine weather.—Hove to at 2 a. m., and at six made sail, and steered W. by S. At half past eight made an island ahead, one of the Kingsmill group. Stood in with the land and received a number of canoes alongside, the natives in them however having nothing to sell us but a few beads of their own manufacture. We saw some cocoanut, and other trees upon the shore, and discovered many of the natives upon the beach, and some dogs. The principal food of these islanders is, a kind of breadfruit, which they pound very fine and mix it with fish.

Feb. 8th. Commences squally with fresh breezes from the northward.—Took a departure from Kingsmill Island; one of the group of that name, in Lat. 1° 27' N. and Long. 175° 14' E. In the morning passed through the channel between Marshall's and Gilbert's Islands; luffèd to and despatched a boat to Marshall's Island, but did not land, as the natives appeared hostile, and those who swam off to the boat, endeavoured to steal from her. When about to leave, a volley of musketry was discharged at them, which probably killed or wounded some of them.

The boat then gave chase to a canoe, paddled by two of the natives, which were fired upon when within gunshot, when they immediately ceased paddling; and on the boat approaching them, discovered that one of the natives was wounded. In the most supplicating manner they held up a jacket, manufactured from a kind of flag, and some beads, being all they possessed, giving their inhuman pursuers to understand, that all should be theirs if they would spare their lives! The wounded native laid down in the bottom of the boat, and from his convulsed frame and trembling lip, no doubt remained but that the wound was mortal.

The boat then returned on board and we made sail for the Mulgrave Islands. Here was another sacrifice; an innocent child of nature shot down, merely to gratify the most wanton and unprovoked cruelty, which could possibly possess the heart of man. The unpolished savage, a stranger to the more tender sympathies of the human heart, which are cultivated and enjoyed by civilized nations, nurtures in his bosom a flame of revenge, which only the blood of those who have injured him, can damp; and when years have rolled away, this act of cruelty will be remembered by these islanders, and made the pretext to slaughter every white man who may fall into their hands.

Feb. 11th. Commenced with strong breezes from the Northward. At half past meridian made the land bearing E. N. E. four leagues distant. Stood in and received a number of canoes alongside. Sent a boat onshore; and brought off a number of women, a large quantity of cocoanuts, and some fish.—Stood offshore most of the night, and Feb. 12th, in the morning stood inshore again and landed the women.—We then stood along shore looking out for an anchorage, and reconnoitring the country, in the hope of finding some spot suitable for cultivation; but in this we were disappointed, or more properly speaking, they, the mutineers; for we had no will of our own, while our bosoms were torn with the most conflicting passions, in which Hope and Despair alternately gained the ascendency.

Feb. 13th. After having stood off all night, we in the morning stood in, and after coasting the shores of several small islands, we came to one, low and narrow, where it was determined the Ship should be anchored. When nearly ready to let go, a man was sent into the chains to sound, who pronounced twelve fathoms; but at the next cast, could not get bottom. We continued to stand in, until we got regular sounding, and anchored within five rods of the shore, on a coral rock bottom, in seven fathoms of water. The ship was then moored with a kedge astern, sails furled, and all hands retired to rest, except an anchor watch.

Feb. 14th, was spent in looking for a landing place. In the morning a boat was sent to the Eastward, but returned with the information that no good landing place could be found, the shore being very rocky. At 2 p. m. she was sent in an opposite direction, but returned at night without having met with better success; when it was determined to land at the place where we lay; notwithstanding it was very rocky.—Nothing of consequence was done, until Sunday, 15th Feb. 1824, when all hands were set to work to construct a raft out of the spare spars, upon which to convey the provisions, &c. on shore.

The laws by which we were now governed had been made by Comstock, soon after the mutiny, and read as follows:

> That if anyone saw a sail and did not report it immediately, he should be put to death! If anyone refused to fight a ship he should be put to death; and the manner of their death, this—

They shall be bound hand and foot and boiled in the try pots, of boiling oil!

Every man was made to seal and sign this instrument, the seals of the mutineers being black, and the remainder, blue and white. The raft or stage being completed, it was anchored, so that one end rested upon the rocks, the other being kept seaward by the anchor. During the first day many articles were brought from the ship in boats, to the raft, and from thence conveyed on shore. Another raft, however, was made, by laying spars upon two boats, and boards again upon them, which at high water would float well up on the shore. The following, as near as can be recollected, were the articles landed from the ship; (and the intention was, when all should have been got on shore, to haul the ship on shore, or as near it as possible and burn her.)

One mainsail, one foresail, one mizen-topsail, one spanker, one driver, one maintop gallant-sail, two lower studdingsails, two royals, two topmast-studdingsails, two top-gallant-studdingsails, one mizen-staysail, two mizen-top-gallant-sails, one fly-gib, (thrown overboard, being a little torn,) three boat's sails (new,) three or four casks of bread, eight or ten barrels of flour, forty barrels of beef and pork, three or more 60 gal. casks of molasses, one and a half barrels of sugar, one barrel dried apples, one cask vinegar, two casks of rum, one or two barrels domestic coffee, one keg W. I. coffee, one and a half chests of tea, one barrel of pickles, one barrel cranberries, one box chocolate, one cask of tow-lines, three or more coils of cordage, one coil rattling, one coil lance warp, ten or fifteen balls spunyarn, one ball worming, one stream cable, one larboard bower anchor, all the spare spars, every chest of clothing, most of the ship's tools, &c. &c. The ship by this time was considerably unrigged.

On the following day, Monday 16th February, Payne the second in the mutiny, who was on board the ship attending to the discharge of articles from her, sent word to Comstock, who with Gilbert Smith and a number of the crew were on shore, attending to the landing of the raft;

> That if he did not act differently with regard to the plunder, such as making presents to the natives of the officers' fine clothing, &c. he would do no more, but quit the ship and come on shore.

Comstock had been very liberal to the natives in this way, and his object was, no doubt, to attach them as much as possible to his person, as it must have been suggested to his guilty mind, that however he himself might have become a misanthrope, yet there were those around him, whose souls shuddered at the idea of being forever exiled from their country and friends, whose hands were yet unstained by blood, but who might yet imbrue them, for the purpose of escape from lonely exile, and cruel tyranny.

When the foregoing message was received from Payne, Comstock commanded his presence immediately on shore, and interrogated him, as to what he meant by sending such a message. After considerable altercation, which took place in the tent, Comstock was heard to say, "I helped to take the ship, and have navigated her to this place.—I have also done all I could to get the sails and rigging on shore, and now you may do what you please with her; but if any man wants anything of me, I'll take a musket with him!"

"That is what I want," replied Payne, "and am ready!" This was a check upon the murderer, who had now the offer of becoming a duellist; and he only answered by saying, "I will go on board once more, and then you may do as you please."

He then went on board, and after destroying the paper upon which were recorded the "Laws," returned, went into the tent with Payne, and putting a sword into a scabbard, exclaimed, "this shall stand by me as long as I live."

We ought not to omit to mention that during the time he was on board the ship, he challenged the persons there, to fight him, and as he was leaving, exclaimed "I am going to leave you; Look out for yourselves!"

After obtaining from Payne permission to carry with him a cutlass, a knife, and some hooks and lines, he took his departure, and as was afterwards ascertained, immediately joined a gang of natives, and endeavoured to excite them to slay Payne and his companions! At dusk of this day he passed the tent, accompanied by about fifty of the natives, in a direction of their village, upwards of a league distant. Payne came on board, and after expressing apprehensions that Comstock would persuade the natives to kill us all, picked out a number of the crew to go on shore for the night, and stationed sentinels around the tent, with orders to shoot anyone, who should attempt to approach without giving the countersign. The night, however, passed, without any one's appearing; but early on the morning of the—

17th Feb. Comstock was discovered at some distance coming towards the tent. It had been before proposed to Smith by Payne, to shoot him; but poor Smith like ourselves, dare do no other than remain upon the side of neutrality.

Oliver, whom the reader will recollect as one of the wretches concerned in the mutiny, hurried on shore, and with Payne and others, made preparations to put him to death. After loading a number of muskets they stationed themselves in front of the tent, and waited his approach—a bushy spot of ground intervening, he did not make his appearance until within a short distance of the tent, which, as soon as he saw, drew his sword and walked quick towards it, in a menacing manner; but as soon as he saw a number of the muskets levelled at him, he waved his hand, and cried out, "don't shoot me, don't shoot me! I will not hurt you!"

At this moment they fired, and he fell!—Payne fearing he might pretend to be shot, ran to him with an axe, and nearly severed his head from his body! There were four muskets fired at him, but only two balls took effect, one entered his right breast, and passed out near the back bone, the other through his head.

Thus ended the life, of perhaps as cruel, blood-thirsty, and vindictive a being as ever bore the form of humanity.

All hands were now called to attend his burial, which was conducted in the same inconsistent manner which had marked the proceedings of the actors in this tragedy. While some were engaged in sewing the body in a piece of canvas, others were employed in digging a grave in the sand, adjacent to the place of his decease, which, by order of Payne, was made five feet deep. Every article attached to him, including his cutlass, was buried with him, except his watch; and the ceremonies consisted in reading a chapter from the bible over him, and firing a musket!

Only twenty-two days had elapsed after the perpetration of the massacre on board the ship, when with all his sins upon his head, he was hurried into eternity!

No duty was done during the remainder of the day, except the selection by Payne, of six men, to go on board the ship and take charge of her, under the command of Smith; who had communicated his intentions to a number of running away with the ship. We think we cannot do better than to give an account of their escape in the words of Smith himself. It may be well to remark, that Payne had ordered the two binnacle compasses to be brought on shore, they being the only ones remaining on board, except a hanging compass suspended in the cabin. Secreting one of the binnacle compasses, he took the hanging compass on shore, and the exchange was not discovered.

At 7 p. m. we began to make preparations for our escape with the ship.—I went below to prepare some weapons for our defence should we be attacked by Payne, while the others, as silently as possible, were employed in clearing the running rigging, for everything was in the utmost confusion. Having found one musket, three bayonets, and some whale lances, they were laid handy, to prevent the ship being boarded. A handsaw well greased was laid upon the windlass to saw off the cable, and the only remaining hatchet on board, was placed by the mizen mast, to cut the stern moorings when the ship should have sufficiently swung off.

Taking one man with me, we went upon the fore-top-sail-yard, loosed the sail and turned out the reefs, while two others were loosing the main-top-sail and main sail. I will not insult the reader's good sense, by assuring him, that this was a duty, upon the success of which seemed to hang our very existence. By this time the moon was rising, which rendered it dangerous to delay, for those who had formed a resolution to swim on board, and accompany us. The bunts of the sails being yet confined aloft, by their respective gaskets, I sent a man on the fore-yard and another upon the fore-top-sail-yard, with orders to let fall, when I should give the word; one man being at the helm, and two others at the fore tack.

It was now half past nine o'clock, when I took the handsaw, and in less than two minutes the cable was off!—The ship payed off very quick, and when her head was off the land, there being a breeze from that quarter, the hawser was cut and all the sail we could make upon the ship immediately set, a fine fair wind blowing. A raft of iron hoops, which was towing alongside, was cut adrift, and we congratulated each other upon our fortunate escape; for even with a vast extent of ocean to traverse, hope excited in our bosoms a belief that we should again embrace our friends, and our joy was heightened by the reflection, that we might be the means of rescuing the innocents left behind, and having the guilty punished.

After a long and boisterous passage the ship arrived at Valparaiso, when she was taken possession of by the American Consul, Michael

Hogan, Esq. and the persons on board were put in irons on board a French frigate, there being no American man-of-war in port. Their names were, Gilbert Smith, George Comstock, Stephen Kidder, Joseph Thomas, Peter C. Kidder, and Anthony Henson.

Subsequently they were all examined before the U. S. Consul; and with the following, an examination of Gilbert Smith, we shall commence another chapter.

CHAPTER 3

Surviving the Massacre

U. S. Consulate,
Valparaiso, 15th June, 1824.
Gilbert Smith examined on oath, touching the mutiny and murder on board the whale ship *Globe*, of Nantucket, Massachusetts, in the Pacific Ocean.

Question. Who were the Captain and mates of the ship *Globe*?

Ans. Thomas Worth, Captain; William Beetle, first mate; John Lumbert, second mate; Nathaniel Fisher, third mate.

Q. Where was you born?

A. In the town of Edgarton, State of Massachusetts.

Q. Did you sail from thence in the ship *Globe* of Nantucket, 20th Dec. 1822, and in what capacity?

A. Yes; as a boat-steerer.

Q. Was there anything like mutiny on board the ship during her passage to the Sandwich Islands?

A. No.

Q. How many men belonged to the ship on sailing from Nantucket?

A. Twenty-one in all.

Q. Did any run away at the Sandwich Islands?

A. Six men ran away, and one was discharged.

Q. How many men were shipped in their places?

A. John Oliver, of Shields, England; Silas Payne, of Rhode Island; Thomas Lilliston, of Virginia; William Steward, of Philadelphia, (black;) Anthony Henson, of Barnstable; and a native of the Sandwich Islands.

Q. On what day or night did this murderous mutiny take place?

A. On Sunday night the 26th of January, this year; in the morning of that day there was a great disturbance, in consequence of Joseph Thomas having insulted the Captain, for which he was whipped by the Captain, with the end of the main buntline. The part of the crew not stationed stood in the hatchway during the punishment.

Q. Did anything happen in consequence, during that day?

A. No: I lived aft; I heard nothing about it; Capt. Joy of the *Lyra*, was on board nearly all day.

Q. How were you stationed during the night?

A. The Captain, first and second mates, kept no watch during that night; the rest of the crew were stationed in three watches, in charge of the third mate and boat-steerers.

Q. Who had charge of the first watch during that night?

A. I had charge of the watch from 7 to 10 o'clock. At 8 the Captain came on deck, and had two reefs taken in the topsails, and at 9 went down, leaving me the orders for the night, to keep the ship by the wind, until two o'clock, and not to tack until the other watch came up; and on tacking, a light to be set for the *Lyra* who was in company, to tack also.

At 10 o'clock I went below, being relieved by the boat-steerer Comstock, to whom I passed the orders given me by the Captain,——(*Here follows a detailed account of the mutiny, with which the reader has already been made acquainted.*)

Q. Do you believe that Joseph Thomas had any knowledge of Comstock's intent to commit murder that night?

A. I think he must have known something about it, according to his talk.

Q. Do you believe that any other person in the ship, besides those persons who committed the murder, knew of the intention?

A. Thomas Lilliston knew about it, because he went to the cabin door with an axe, and a boat knife in his hand, in company with the murderers, but he did not go below.

Q. Did you live with them aft, afterwards?

A. No: I lived in the forecastle, but all on board eat in the cabin.

Q. Name all the persons you left on the island, where you cut the cable of the ship and escaped.

A. Silas Payne, John Oliver, (being the principal mutineers next to Samuel B. Comstock,) Thomas Lilliston, Rowland Coffin, William Lay, Cyrus M. Hussey, Columbus Worth, Rowland Jones, and the Sandwich Island native, called Joseph Brown. The last five I believe ignorant of any knowledge of the intent to murder.

Q. What became of Samuel B. Comstock, who was the head mutineer after he landed upon the island?

A. He was shot on the morning of the 17th Feb. by Silas Payne, and John Oliver, his associates in all the mutiny and murderous course they had pursued, and buried five feet deep on the beach near their tent; a chapter was read from the bible by me, acting under the orders of Payne, and muskets were fired by his orders, by the men.

Q. Why did they murder Comstock?

A. For giving away to the natives clothes and other articles before they were divided.

Q. Were the natives friendly and quiet?

A. Yes; very peaceable, gave away any thing they had; bread fruit, cocoanuts and other things.

Q. How did Joseph Thomas conduct himself during the passage from the isle to this port?

A. In common, when help was called, he was the first man disobedient, and frequently said he would do as he pleased.

Q. Did he often speak of the murder, or of his knowing it about to take place?

A. I only remember, having heard him twice. I told him when we arrived, I would inform the American Consul of it; to which he replied, he should own all he knew about it.

Q. To what State does he belong to your knowledge?

A. To the State of Connecticut, he says.

(Signed) Gilbert Smith.

Sworn to, before me at Valparaiso, this eighteenth day of June, 1824.

(Signed) Michael Hogan,
U. S. Consul.

The examination of the others who came in the ship, was but a repetition of the foregoing. All, however, concurred in believing, that Joseph Thomas was privy to the intention to mutiny, and murder the officers.

The ship was then furnished with necessary sails and rigging, and placed in charge of a Captain King, who brought her to the island of Nantucket, arriving on Sunday 21st November, 1824. Another examination was held before Josiah Hussey, Esq. and all testified, as before the American Consul at Valparaiso.

Thomas, who was put in irons as soon as the land was discovered, was arraigned before the above named justice, and after an elaborate hearing, the prisoner was committed to jail, to take his trial at the following term of the U. S. District Court, and the witnesses recognised in the sum of three hundred dollars each.

Leaving Thomas, awaiting his trial, and the others in the enjoyment of the society of their families and friends, we will return to the Mulgrave Islands, the scene of no inconsiderable portion of our distresses and adventures.

On the 17th Feb. when night came, the watch was set consisting of two men, whose duty it was to guard against the thefts of the natives. At about 10 p. m. all hands were awakened by the cry; "The ship has gone, the ship has gone!" Everyone hastened to the beach and verified the truth of the report for themselves. Some who were ignorant of the intention of Smith and others, to take the ship, were of opinion that the strong breeze then blowing, had caused her to drag her anchor, and that she would return in the morning.

The morning came, but nothing was to be seen upon the broad expanse of ocean, save here and there a solitary seagull, perched upon the crested billow. Payne in a paroxysm of rage, vented the most dreadful imprecations; swearing that could he get them once more in his power, he would put them to instant death. Not so with us; a ray of hope shot through our minds, that this circumstance might be the means of rescuing us from our lonely situation.—The writers of this narrative were upon the most intimate terms, and frequently, though carefully, sympathized with each other upon their forlorn situation.

We dare not communicate our disaffection to the Government of the two surviving mutineers, (Payne and Oliver,) to the others, fearing they might not agree with us in opinion, and we had too good reason to believe, that there was one, who although unstained by blood, yet from his conduct, seemed to sanction the proceedings of the mutineers.

The natives assembled in great numbers around the tent, expressing great surprise at the ship's having left,—Payne gave them to understand that the wind had forced her to sea, and that from her want of sails, rigging, &c. she must be lost, and would never return.—The natives received the assurance with satisfaction, but it was evident, Payne apprehended her safe arrival at some port, and his own punishment; for we were immediately set to work, to tear one boat to pieces, for the purpose of raising upon another, which was to have a deck; Payne, alleging as a reason for this, that the natives might compel us to leave the island. We leave the reader to judge, however, of his motives, while we proceed to give an account of what actually did transpire.

The natives in considerable numbers continued to attend us, and while the work was progressing, exhibited a great deal of curiosity. Their deportment towards us continued to be of the most friendly nature, continuing to barter with us, giving us breadfruit, cocoanuts, &c. for which they received in return, pieces of iron hoop, nails, and such articles as we could conveniently spare.

The small islands of this group are frequently only separated by what are sometimes denominated causeways, or in other words, connected by reefs of coral, extending from the extreme point of one island and connecting it with another. These reefs are nearly dry at low water, and the communication is easily kept up between them by the natives on foot.

On the 19th, in the morning, having obtained permission, several of us left the tent, travelling to the Eastward.—After crossing upon the causeways to several adjacent islands, we discovered numerous tracks of the natives in the sand, and having followed them about seven miles, came to a village consisting of about twenty or thirty families; and were received by them with great hospitality. They presented us with breadfruit and the milk of cocoanuts, while the wonder and astonishment of those who had not as yet seen us, particularly the women and children, were expressed by the most uncouth grimaces, attended with boisterous laughter, and capering around us. What more particularly excited their astonishment was the whiteness of our skins, and their mirth knew no bounds when they heard us converse.

Early on the morning of the 20th, we were ordered to go to work upon the boat; but at the request of a number, this duty was dispensed with, and we permitted to stroll about the island. A number went to the village, carrying with them muskets, at the report of which and the effect produced by the balls, the natives were struck with wonder and astonishment.

The reader will no doubt agree with us when we pronounce this to have been a bad policy, for they certainly disliked to have visitors possessed of such formidable and destructive weapons. They however continued to visit the tent without discovering any hostile intentions, and we continued to put the utmost confidence in them, or more properly speaking to live without any fear of them.

I (William Lay,) left the tent on a visit to the village, where I was received with the same kindness as before.—An old man between fifty and sixty years of age, pressed me to go to his house and tarry during the night, which I did.—The natives continued in and around the tent until a late hour, gratifying their curiosity by a sight of me. I was provided with some mats to sleep upon, but the rats, with which the island abounds, prevented my enjoying much sleep.

At 10 o'clock I took my leave of them, with the exception of a number, who accompanied me to the tent.

Silas Payne and John Oliver, together with two or three others, set out in one of the boats, for the purpose of exploring the island, and making new discoveries, leaving the rest of us to guard the tent. They were absent but one night, when they returned, bringing with them two young women, whom Payne and Oliver took as their wives. The women apparently showing no dissatisfaction, but on the contrary appeared much diverted. Payne now put such confidence in the natives, that he dispensed with having a watch kept during the night, and slept as secure as though he had been in his native country.

Payne, on awaking near morning, found the woman that he had brought to live with him was missing. After searching the tent, and finding nothing of her, concluded she had fled. He accordingly armed himself, together with John Oliver and Thomas Lilliston, (with muskets,) and set out for the nearest village, for the purpose of searching her out. They arrived at the village before it was light, and secreted themselves near an Indian hut, where they awaited the approach of day, in hopes of seeing her.

Accordingly at the approach of daylight, they discovered the hut literally thronged with natives, and among the number, they discovered the woman they were in search of. At this moment one of them fired a blank cartridge over their heads, and then presented themselves to their view, which frightened the natives in such a manner that they left the hut and fled. Payne then pursued after, firing over their heads till he caught the one he wanted, and then left the village for his own tent.—On arriving at the tent, he took her, gave her a severe flogging and then put her in irons, and carried on in this kind of style until he was by them killed, and called to render up his accounts to his offended Judge.

This severity on the part of Payne, irritated the natives, and was undoubtedly the cause of their committing depredations and theft, and finally murdering all our remaining crew, excepting myself and Hussey.

Early on the succeeding morning, it was discovered that the tool chest had been broken open, and a hatchet, chisel, and some other articles, purloined by the natives. Payne worked himself into a passion, and said he would be revenged. During the day he informed a number of the natives of what had been done, (who signified much regret at the circumstance,) and vowing vengeance if the articles were not returned. During this day the natives frequented the tent more than they had ever done before; and at night one of them came running with one half of the chisel which had been stolen, it having been broken in two.

Payne told them it was but half of what he required, and put the Indian in irons, signifying to him, that in the morning he must go with him to the village, and produce the rest of the articles, and also point out the persons engaged in breaking open the chest. The poor native seemed much chagrined at his confinement; yet his companions who remained near the tent during the night, manifested no dissatisfaction, which we could observe.

In the morning, Payne selected four men, *viz*: Rowland Coffin, Rowland Jones, Cyrus M. Hussey, and Thomas Lilliston, giving them each a musket, some powder and fine shot; declining to give them balls, saying, the report of the muskets would be sufficient to intimidate them. The prisoner was placed in charge of these men, who had orders to go to the village, and recover the hatchet and bring back the person whom the prisoner might point out as the thief.

They succeeded in getting the hatchet, but when about to return, the natives in a great body, attacked them with stones. Finding that they retreated, the natives pursued them, and having overtaken Rowland Jones, killed him upon the spot. The remainder, although bruised with the stones which these islanders had thrown with great precision, arrived at the tent with the alarming intelligence of a difficulty;—while they followed in the rear armed for war!

No time was lost in arming ourselves, while the natives collected from all quarters, and at a short distance from the tent, seemed to hold a kind of council. After deliberating some time, they began to tear to pieces one of the boats.

These were of vital importance to our guilty commander, and he ventured to go to them for the purpose of pacifying them. One of the chiefs sat down upon the ground with him, and after they had sat a few moments, Payne accompanied the chief into the midst of the natives. After a conference with them which lasted nearly an hour, he returned to the tent, saying that he had pacified the natives upon the following conditions. They were to have every article belonging to us, even to the tent; and Payne had assured them of his willingness, and that of the others to live with, and be governed by them, and to adopt their mode of living!

We have reason to doubt the sincerity of Payne in this respect, for what was to us a hope which we cherished with peculiar pleasure, must have been to him, a source of fearful anticipation—we mean the probable safe arrival of the ship, in the U. S. which should result in our deliverance. Our situation at this time was truly alarming; and may we not with propriety say, distressing?

Surrounded by a horde of savages, brandishing their war clubs and javelins, our more than savage commanders, (Payne and Oliver) in anxious suspense as to the result of their negotiations with them; no refuge from either foe, and what contributed not a little to our unhappiness, was a consciousness of being innocent of having in the least manner wilfully aided the destroyers of the lives of our officers, and the authors of our now, truly unhappy situation.

The natives now began to help themselves to whatever articles suited them, and when some of them began to pull the tent down, an old man and his wife took hold of me, and after conducting me a few rods from the tent, sat down, keeping fast hold of my hands. Under the most fearful apprehensions I endeavoured to get from them, but they insisted upon detaining me. I endeavoured to console myself with the idea, that gratitude had prompted them to take care of me, as I had frequently taken the part of this old woman, when she had been teased by others; but alas! the reflection followed, that if this was the case, there was a probability that not only my bosom friend, was about to be sacrificed, but I should be left alone to drag out a weary existence, with beings, strangers to the endearing ties which bind the hearts of civilized man.

Whether Payne and his associates offered any resistance to the course now pursued by the natives or not, I do not know. Suffice it to say, that all at once my ears were astounded with the most terrifying whoops and yells; when a massacre commenced but little exceeded by the one perpetrated on board the *Globe*. Our men fled in all directions, but met a foe at every turn. Lilliston and Joe Brown (the Sandwich Islander,) fell within six feet of me, and as soon as down, the natives macerated their heads with large stones. The first whom I saw killed, was Columbus Worth. An old woman, apparently sixty years of age, ran him through with a spear, and finished him with stones!

My protectors, for now they were truly so, shut out the scene by laying down upon the top of me, to hide me from the view of the merciless foe! I was however discovered, and one of the natives attempted to get a blow at me with a handspike, which was prevented by them; when, after a few words, he hurried away.

As soon as the work of death had been completed, the old man took me by the hand and hurried me along towards the village. My feet were very much lacerated in passing over the causeways of sharp coral rock, but my conductor fearing we might be pursued, hurried me onward to the village, where we arrived about noon. In a few minutes the *wigwam* or hut of the old man, was surrounded, and all seeming to talk at once, and with great excitement, I anticipated death every moment. Believing myself the sole survivor, the reader must pardon any attempt to describe my feelings, when I saw a number of the natives approaching the hut, and in the midst, Cyrus M. Hussey, conducted with great apparent kindness.

Notwithstanding we had both been preserved much after the same manner, we could not divest ourselves of the apprehension, that we perhaps had been preserved, for a short time, to suffer some lingering death.

Our interview was only long enough to satisfy each other that we alone survived the massacre, when we were separated; Hussey being taken away, and it seemed quite uncertain, even if our lives were spared, whether we ever saw each other again.

CHAPTER 4

After the Massacre

On the following day, however, accompanied by natives, we met at the scene of destruction, and truly it was an appalling one to us. The mangled corpses of our companions, rendered more ghastly from the numerous wounds they had received, the provisions, clothing, &c. scattered about the ground, the hideous yells of exultation uttered by the natives, all conspired to render our situation superlatively miserable.

We asked, and obtained leave from our masters, to bury the bodies which lay scattered about. We dug some graves in the sand, and after finishing this melancholy duty, were directed to launch the canoes, preparatory to our departure, (for we had come in canoes) when we begged permission, which was readily granted, to take some flour, bread and pork, and our respective masters assisted us in getting a small quantity of these articles into the largest canoe. We also took a blanket each, some shoes, a number of books, including a bible, and soon arrived at the landing place near the village.

As the natives seemed desirous of keeping us apart, we dare not make any inquiries for each other, but at my request, having boiled some pork in a large shell, Hussey was sent for, and we had a meal together; during which time, the natives assembled in great numbers, all anxious to get a sight, not only of our novel mode of cutting the meat and eating it, but of the manner in which we prepared it. One of them brought us some water in a tin cup, as they had seen us drink frequently when eating.

The natives now began to arrive from distant parts of the islands, many of whom had not yet heard of us, and we were continually subjected to the examination of men, women and children. The singular colour of our skin, was the greatest source of their admiration, and we were frequently importuned to adopt their dress.

On the 28th Feb. early in the morning the whole village appeared to be in motion. All the adults commenced ornamenting themselves, which to me appeared to render them hideous. After greasing themselves with coconut oil, and hanging about them numerous strings of beads, they set off, taking us with them, to a flat piece of ground, about half a mile distant, where we found collected a great number, and all ornamented in the same fantastic manner.—Knowing that many of the natives inhabiting islands in the Pacific Ocean, are cannibals, we were not without our fears that we had been preserved to grace a feast!

Our apprehensions, however, were dissipated, when we saw them commence a dance, of which we will endeavour to give the reader some idea. The only musical instrument we saw, was a rude kind of drum; and the choristers were all females, say twenty or thirty, each having one of these drums. The music commenced with the women, who began upon a very low key, gradually raising the notes, while the natives accompanied them with the most uncouth gesticulations and grimaces. The precision with which about three hundred of these people, all dancing at a time, regulated their movements, was truly astonishing; while the yelling of the whole body, each trying to exceed the other, rendered the scene to us, not only novel, but terrific.

The dance ended near night, and those natives who lived in a distant part of the island, after gratifying their curiosity by gazing upon us, and even feeling of our skins, took their departure.

After our return to the village, we cooked some meat upon the coals, and with some bread, made a hearty meal. One source of regret to us, was, that the natives began to like our bread, which heretofore they had scarcely dared to taste; and particularly the woman whom I called mistress, ate, to use a sea phrase, her full allowance.

The natives expressed great dislike at our conversing together, and prohibited our reading, as much as possible. We never could make them comprehend that the book conveyed ideas to us, expressed in our own language.

Whether from a fear that we might concert some plan of escape, or that we might be the means of doing them some injury while together, we know not;—but about the first of April, we discovered that we were about to be separated! The reader may form some idea of our feelings when we were informed that Hussey was to be taken by his master and family, to a distant part of the island! Not having as yet become sufficiently acquainted with their language, we were unable to comprehend the distance from our present location.

It now becomes expedient to present the reader with our separate accounts, in which we hope to be able to convey an idea of the manners and customs of these people. We had experienced in a very short time so many vicissitudes, and passed through so many scenes of distress, that no opportunity was afforded to keep a journal, and notwithstanding we had even lost the day of the week and month, yet with such force, were the principal incidents which occurred during our exile, impressed upon our minds, that we can with confidence proceed with our narrative, and will commence the next chapter with an account of the adventures of William Lay.

Chapter 5

Adventures of William Lay

Early in the morning of the day on which Hussey left me, preparations were made for his embarkation with his new master and family. We were allowed a short interview, and after taking an affectionate leave of each other, we parted with heavy hearts. The tender ties which bound me to my companion in misfortune, seemed now about to be forever broken asunder. No features to gaze upon, but those of my savage masters, and no one with whom I could hold converse, my heart seemed bursting with grief at my lonely situation.—On the departure of my companion, the "star of hope" which had often gleamed brightly mid the night of our miseries, seemed now about to set forever!

After watching the canoe which bore him from me, until she was hid from my view in the distance, I returned to the hut with my master, and as I had eaten but little during the day, the calls of nature induced me to broil my last morsel of meat, with which, and some bread, I made a tolerable supper. The natives began to be very fond of the bread, and eat of it as long as it lasted, which unfortunately for me, was but a short time.

I informed my master that I should like to have some more of the meat from the place where the ship had lain. On the following morning, my master, mistress, and four or five others embarked in a canoe, to assist me in procuring some provisions. Observing that they carried with them a number of clubs, and each a spear, I was apprehensive of some design upon my own person; but happily, was soon relieved, by seeing them wade round a shoal of fish, and after having frightened them into shoal water, kill a number with their spears.

We then proceeded on, and when we arrived at the tent, they cooked them after the following manner. A large fire was kindled, and after the wood was burned to coals, the fish were thrown on, and snatched and eaten as fast as cooked; although they were kind enough to preserve a share for me, yet the scene around me, prevented my enjoying with them, their meal. The tent which had been torn down, had contained about forty barrels of beef and pork, two hogsheads of molasses, barrels of pickles, all the clothing and stores belonging to the ship, in short, everything valuable, such as charts, nautical instruments, &c. &c.

The latter had been broken and destroyed, to make ornaments, while the beef, pork, molasses and small stores lay scattered promiscuously around. They appeared to set no value upon the clothing, except to tear and destroy it. The pieces of beef and pork, from the barrels, (which had been all stove,) were scattered in every direction, and putrefying in the sun. After putting into the canoe some pork and a few articles of clothing, we commenced our return;—but a strong head wind blowing, we had considerable difficulty in getting back.

For some considerable time, nothing material occurred, and I led as monotonous and lonely a life, as could well be imagined. It is true, I was surrounded by fellow beings; and had all hope of ever seeing my country and friends again, been blasted, it is probable I might have become more reconciled to my condition, but I very much doubt if ever perfectly so, as long as reason and reflection held their empire over my mind. My books having been destroyed from a superstitious notion of their possessing some supernatural power, I was left to brood over my situation unpitied and alone.

Sometime in July, as I judged, Luckiair, son-in-law to my master, Ludjuan, came from a distant part of the group, on a visit, and during the week he remained with us, we became much attached to each other. When he told me, that on his return he should pass near the place where Hussey lived, my anxiety to accompany him thus far, was so great, that after much persuasion, Ludjuan gave his consent for me to go. On our way we stopped at the tent, and I procured for the last time, a small quantity of the ship's provisions, although the meat was some of it in a very decayed state.

In consequence of head winds, we were compelled to stop for the night upon a small island, where we found an uninhabited hut; and after cooking some meat, and baking some wet flour (for it was no other) in the ashes, we took our mats into the hut, and remained until next day. The wind continuing to blow fresh ahead, we gathered some green breadfruit, and cooked some meat, in the same manner as they cook the largest of their fish, which is this.—A hole is dug in the ground, and after it has been filled with wood, it is set on fire, and then covered with stones. As the wood burns away, the heated stones fall to the bottom, which, when the fire is out, are covered with a thick layer of green leaves, and then the meat or fish is placed upon these leaves, and covered again in a careful and ingenious manner, and the whole covered with earth. This preserves the juices of the fish, and in this way do they cook most of their fish, with hot stones.

In the afternoon the weather proving more favourable, we left our encampment, and at sun down arrived at a place called Tuckawoa; at which place we were treated with the greatest hospitality. When we were about to leave, we were presented with breadfruit and cocoanuts in great abundance. As we approached the place of Hussey's residence, I discovered him standing on the beach. Our joy at meeting, I will not attempt to describe.—We had a short time, however, allowed us, in which to relate our adventures, and condole with each other; for in an hour we were once more separated; and we pursued our course for the residence of Luckiair.

After encamping another night upon the beach, we at length arrived at the house of my conductor, which was at a place called Dillybun. His family consisted of his wife and one child, whom we found busily engaged in making a fishing net. When near night Luckiair and myself went out and gathered some breadfruit, and after making a hearty meal, slept soundly upon our mats until morning.

A little before noon on the following day, two natives with their wives, arrived from Lujnonewort, the place where Hussey lived, and brought me some flour, and a piece of meat. The natives would eat of the bread, but would not taste of the meat. I remained here about a week, when Ludjuan came for me. Nothing occurred of note, during our passage back to Milly, (the place of my residence,) where I was welcomed by the natives with every demonstration of joy. I was sent for by one of the chiefs, who asked many questions, and as a mark of his friendship for me, when I was about to return, presented me with a kind of food called *chakaka*.

My present consisted of a piece about two feet long and six inches in diameter. It is made of a kind of fruit common among these islands, and called by the inhabitants, *bup*. The fruit is scraped very fine, and then laid in the sun until perfectly dry. Some of the leaves of the tree bearing the fruit, are then wrapped round a piece of wood, which is the mould or former, and when securely tied with strings, the former is withdrawn, and into this cylinder of leaves is put the *bup*, which is of a sweet and pleasant taste.

At the urgent request of the natives, I now adopted their dress. Having but one pair of trowsers and a shirt left, I laid them by for bad weather, and put on the costume of a Mulgrave islander. This dress, if it may be so called, consists in a broad belt fastened round the waist, from which is suspended two broad tassels. The belt is made from the leaves of the *bup* tree, and very ingeniously braided, to which is attached the tassels, which are made of a coarser material, being the bark of a small vine, in their language called *aht-aht*. When the dress is worn, one of the tassels hangs before and the other behind.

The sun, as I expected, burned my skin very much; which the natives could not account for, as nothing of the kind ever happened among themselves.

One day there was seen approaching a number of canoes, which we found were loaded with fish for the chiefs, and to my great joy, Hussey was one of the passengers. My master accompanied me to see him; and we anticipated at least a mental feast in each other's society. But of this enjoyment we were deprived by the natives, who were always uneasy when we were conversing together.

I learned, however, from Hussey, that the natives had been kind to him; but before we had an opportunity to communicate to each other our hopes and fears, he was hurried away. Having now gained considerable knowledge of their language, I learned that they were afraid that if we were permitted to hold converse, we should be the means of provoking the Supreme God, *Anit*, to do them some injury.

The breadfruit beginning to ripen, we were all employed in gathering it; and I will endeavour to give the reader an idea of the process of preserving it. After the fruit was gathered, the outside rind was scraped off, and the seeds taken out; which are in size and appearance like a chesnut. The fruit is then put into a net, the meshes of which are quite small, taken into the salt water, and then beat with a club to pummice. It is then put into baskets made of cocoanut leaves, and in about two days becomes like a rotten apple; after which the cores are taken out, and the remainder after undergoing a process of kneading, is put into a hole in the ground, the bottom and sides of which are neatly inlaid with leaves, and left about two days; when it again undergoes the same process of kneading, and so on, until it becomes perfectly dry.—This occupied us a number of days; and when we were engaged in gathering another, and a larger kind, a small boy came running towards us, and exclaimed, "*Uroit a-ro rayta mony la Wirrum*," that is, "the chiefs are going to kill William." Ludjuan seeing that I understood what the boy said, he said "*reab-reab!*" "it is false."

From the pains taken by the natives to keep Hussey and myself apart, it was evident that they were in some measure afraid of us; but from what cause I had yet to learn. After passing a sleepless night, we again in the morning pursued our labours, but I was continually agitated by fearful apprehensions. About midnight I overheard some of the natives in the tent talking about me, and I was now convinced that some injury was contemplated. I then asked them what I was to be killed for. They seemed surprised when I told them I had been listening; yet they denied that I was to be killed, and one of them who had frequently manifested for me much friendship, came to my mat, and lay down with me, assuring me I should not be injured.

The harvest being ended, a feast was had, and the chiefs were presented with considerable quantities of this fruit, after it had been prepared and baked, which in taste resembled a sweet potato, sending presents of it in all directions about the island.

Having now but little work to do, I confined myself to the hut as much as possible, for I had been observed for some time in a very suspicious manner. In a few days I was informed that Hussey had been brought to the island, and it was immediately suggested to my anxious mind, that we were now to be sacrificed. Ludjuan went with me to see Hussey, but we were only allowed a few moments conversation, when I was taken back to the hut, and communicated my fears to my old mistress, who sympathized with me, but said if the chiefs had determined it, there was no hope for me. I now was made acquainted with the cause of their dislike, which was no less than a superstitious idea, that we were the cause of a malady, then raging to considerable extent!

This disease consisted in the swelling of the hands and feet, and in many instances the faces of the youth swelled to such a degree, that they were blind for a number of days. Such a disease they had never before been afflicted with. I had now an opportunity of most solemnly protesting my total inability to injure them in this way, and as the disease had as yet caused no death, I had a hope of being spared. I learned that a majority of the chiefs in council, were for putting me to death, but one of them in particular, protested against it, fearing it might be the cause of some worse calamity. As the vote to carry into effect any great measure, must be unanimous, this chief was the means by his dissenting, of saving my life.

The afflicted began to recover, and my fears were greatly lessened; but as these people are of a very unstable and changeful character, I could not entirely divest myself of apprehensions.

As soon as the harvest was completed, great preparations were made for the embarkation of the chiefs, who were going to make their annual visit to the different islands. They told me that the King, whom they called Laboowoole, yet, lived on an island at the N.W. and if he did not receive his yearly present of preserved breadfruit and *pero*, he would come with a great party to fight them. Twelve canoes were put in the water, each one carrying a part of the provisions, and manned by about two hundred persons.

After an absence of four or five days, during which time we exchanged civilities with numerous chiefs, we returned to Milly, and hauled up the canoes. I now learned that the principal chief, had said that it would have been wrong to kill me, firmly believing that the disease with which they had been afflicted, had been sent by their God, as a punishment for having killed Payne and the others! The malady having now entirely disappeared, they considered that crime as expiated!

About two days after my return, there was great excitement, in consequence of the appearance of a ship! Seeing the natives were very much displeased at the circumstance, I concealed as well as I could, the gladdening emotions which filled my breast; and, surrounded by about three hundred of them, went round a point of land, when I distinctly saw a ship standing for the land. The displeasure of the natives increased, they demanded to know where she came from, how many men she had in her, &c. I was compelled to tell them that she was not coming to get me, and even pretended to be afraid of her approach, which pleased them much, as they appeared determined I should never leave them. At dusk she was so near the land, that I saw them shorten sail, and fondly anticipated the hour of my deliverance as not far distant.

During the night, sleep was a stranger to me, and with the most anxious emotions did I anticipate a welcome reception on board, and above all, a happy and joyful landing on my native shore. In the morning, Ludjuan went with me to the beach, but alas! no ship was in sight. She had vanished, and with her had fled all my hopes of a speedy deliverance. The kind reader can perhaps form some idea of my disappointment.

The natives continued to be kind to me, and I was often complimented by them for my knowledge of their language; and the appearance of my person had very much improved, my hair and beard being long, and my skin turned nearly as black as their own! I was often importuned to have my ears bored and stretched, but never gave my consent, which much surprised them, it being a great mark of beauty. They begin at the age of four years, and perforate the lower part of the ear, with a sharp pointed stick; and as the ear stretches, larger ones are inserted, until it will hang nearly to their shoulders! The larger the ear, the more beauty the person possesses!

About a fortnight after I saw the ship pass, Hussey came with his master, on a visit. His disappointment was great, and we could only cheer each other, by hoping for the best, and wait patiently the pleasure of Heaven.

Hussey again left me, but we parted under less bodings of evil than before, for the kindness of the natives began to increase, and their suspicions to be allayed.

I will here acquaint the reader with some of the means that I was induced to make use of, to satisfy the cravings of appetite. As the island now was in a state of almost entire famine, my daily subsistence not amounting to more (upon an average) than the substance of one half a cocoanut each day. The chief I lived with, having several cocoanut trees that he was very choice of, and which bore plentifully; I would frequently, (after the natives in the hut were all soundly asleep) take the opportunity and get out of the hut unperceived, and climb one of those trees, (being very careful about making the least noise, or letting any of them drop to the ground, whereby I might be detected,) and take the stem of one cocoanut in my mouth, and one in each hand, and in that manner make out to slide down the tree, and would then (with my prize) make the best of my way to a bunch of bushes, at a considerable distance from the hut, where I would have a sumptuous repast; and if any remained, would secrete them, until by hunger, I was drove to the necessity of revisiting that place.

I made a practice of this for some time, until the chief began to miss his cocoanuts, and keep such watch, that I, for fear of being detected, was obliged to relinquish that mode of satisfying my appetite.

A short time after this, I ventured to take a cocoanut off the ground where the natives had recently buried a person; a deed which is strictly against the laws of their religious principles, (if it can be said that they have any,) and a deed which the natives never dare to do, for fear of displeasing their God (*Anit*) under a certain length of time after the person had been buried, and then, the spot is only to be approached by males.

Not twenty-four hours had elapsed after I took the cocoanut, before they missed it, and coming immediately to me, charged me with having taken it, telling me that not a native on the island would have dared so much as to handle it, for fear of the bad spirit, (*Anit*.)

I then told them that I had taken it, but pleading ignorance in the case, and promising never to do anything of the like again, and making it appear to them that I was surprised at what they told me of the bad spirit, and also that I believed the same, they left me, after telling me that if I ever handled another of them, it would not only bring sickness and death upon myself, but would bring it upon the whole island.

The reader will naturally suppose, that my mind was considerably relieved on their leaving me so soon, fearing that something serious might be the result.

After this I was very careful how I did anything that I thought would in the least displease, or irritate them, and made myself content with the portion they saw fit to give me.

I frequently fired a musket to please them, by their request; and told them if they would let me have some powder, I would fire off the swivel, left by the *Globe*. They consented, and collected in great numbers, and after I had loaded the gun with a heavy charge, I told them they had better stand back. They said I must set her on fire, and tell them when she was going off, and they would run! I however, touched her off, when they instantly fell on their faces in the greatest panic. When their fears had subsided, they set up howling and yelling with ecstasy!

They said, if they should have a battle, I must carry that gun with me, which would alone vanquish their enemies!

We were visited by eight or ten canoes, from a distant island, called Alloo. They came to exchange presents with our chiefs, and very soon a great quantity of *pero*, &c. was baked, and having been inspected by the chiefs, to see that it was in a proper state to be presented to their visitors, it was given them to eat.

As these people had never seen me before, I was much annoyed by them. During their stay, I was constantly surrounded; my skin felt of, and often became the sport of the more witty, because my skin was not of so dark a hue as their own, and more especially, as my ears remained in the same form, as when nature gave them to me. These visitors, to my great satisfaction, did not remain long with us.

Their mode of anchoring their canoes is singular. One of them takes the end of a line, and diving to the bottom, secures it to a rock; and in the same way do they dive down to cast it off. I have seen them do this in five fathoms of water.

Chapter 6

Escape From the Island

It was not until the 23rd of December, 1825, that the prospects of being relieved from my disagreeable situation began to brighten. Early in the morning of that day, I was awakened by a hooting and yelling of the natives, who said, a vessel had anchored at the head of the island. They seemed alarmed, and I need not assure the reader, that my feelings were of a contrary nature. Their God was immediately consulted, as to the measures to pursue; but as I was not allowed to be present when he was invoked, I cannot say what was the form of this ceremony, except that cocoanut leaves were used.

Their God, however, approved the plan, which was, that they should go to the vessel, or near her, and swim on board, a few at a time, until two hundred were on board, and then a signal was to be given, when they were to throw the persons on board into the water, and kill them. Two large canoes which would carry fifty men each, were put in readiness, but at first they refused to let me accompany them, fearing that I would inform of their having killed our men, and they would be punished. I assured them that the vessel, having but two masts, did not belong to my nation, and I was certain I could not speak their language.

They at length consented for me to go. We arrived within a few miles of the vessel at night, and early the following morning, were joined by a number of canoes, which made in all two hundred men. It being squally in the forenoon, we remained where we were, but when it cleared up, the yells of the Indians announced the approach of the vessel. I had only time to see that it was really an armed schooner, when I was secreted with their women, about forty in number, in a hut near the shore, and the women had orders to watch me close, that I did not get away.

A boat at this time from the schooner, was seen approaching the shore. She landed at about a hundred yards distant from where I was confined; but it being near night, I soon found she was making the best of her way towards the schooner. Night came, and I was sent for by the principal chief, and questioned closely concerning the schooner. My fears and apprehensions were now excited to a degree beyond human expression, and the kind reader will pardon all attempts to express them.

The natives seeing the whites so bold, excited in them a fear which induced them to flee the island. Accordingly, about midnight, the canoes were launched, and I was carried to a remote part of the island, a distance of about forty miles, where I remained until my fortunate escape.

29th. Early in the morning, we discovered a boat under sail, standing directly for the place where we were; the natives were considerably agitated with fear, and engaged in planning some method by which to overcome the people in the boat, if they should come where we were; and, as I expected, the natives would hide me, as they had heretofore done, I thought it best to offer my services to assist them—I said I would aid them in fighting the boat's crew—and that, as I could talk with them, I would go to them, in advance of the natives, deceive the crew, and prevail on them to come on shore and sit down, and for us to appear friendly till in possession of their arms, then rise upon the crew and kill them without difficulty or hazard.

Some of the natives suspected that I should revolt to the other party, and turn the current of destruction on them; but the chief Luttnon said he liked my plan much, and would inquire of their God, and if he found that I should be true to them, my plan should be adopted. The inquiry resulted in favour of my plan, and they said I might go. The boat was now within one hundred rods of the shore, and Luttnon called me to him, oiled my head and body with coconut oil, and gave me my charge how to conduct. I pledged myself to obey his orders.

My joy at this moment was great, as the boat anchored near where we were. I went to the beach, accompanied by about one hundred of the smartest natives, whom I charged not to manifest a hostile appearance. I hailed the boat in English, and told the crew what the calculations of the natives were, and not to land unless they were well armed. The officer of the boat replied that he would be among them directly; and in a few minutes they landed, (thirteen men and two officers,) and when within a rod of us, I ran to Lieutenant H. Paulding, who took me by the hand, asked if I was one of the *Globe*'s crew, and inquired my name, &c. &c. We then retreated to the boat, facing the natives, who all kept their seats, excepting the one I called father, who came down among us, and took hold of me to carry me back, but desisted on having a pistol presented to his breast.

Lieutenant Hiram Paulding, of the Navy, for such was the name of this gentlemanly officer, informed me that the vessel, was the U. S. Schooner *Dolphin*, sent on purpose to rescue us, and commanded by Lieutenant Com. John Percival.

After expressing my gratitude as well as I was able, to Heaven, which had furnished the means of my deliverance, I acquainted Mr. Paulding, that the only survivor of the *Globe*, except myself, was Cyrus M. Hussey; who was held in bondage upon a neighbouring island. After the boat's crew had taken some refreshment, we left the landing place, and soon arrived at the place where Hussey lived. The natives had concealed him, but after some threatenings from us, restored him, and we were received on board of the *Dolphin*, and treated in the most kind and hospitable manner.

Our hair was now cut, and we were shaved. Our appearance must have been truly ludicrous, our hair having been growing twenty-two months, untouched by the razor or scissors.

Our joy and happiness on finding ourselves on board an American man-of-war, and seeing "the star spangled banner," once more floating in the air, we will not attempt to describe. Suffice it to say, that none can form a true estimate of our feelings, except it be those who have been suddenly and unexpectedly rescued from pain and peril, and threatening death.

In the afternoon the captain wished me to go on shore with him, as an interpreter. We accordingly went, and passed over to the village on the other side of the island, where we had an interview with a woman of distinction, (the men having fled, being principally absent with the chiefs at Alloo.) The captain informed her he wished to see the chiefs, and requested her to send for them that night, that he might visit them in the morning, and make them some presents.

We then returned to the vessel; and the following day, Dec. 1st, went on shore for the purpose of seeing the chiefs, but could not obtain an interview with them. The captain informed the natives that he must see the chiefs, and that he would wait another day, but if disappointed then, he should be compelled to use coercive means. They immediately sent another messenger after them, and we returned on board, accompanied by several of the natives, among whom was Ludjuan. The captain made him several presents, and informed him they were given as a compensation for saving my life. Shortly after, the natives went on shore.

The next morning, Dec. 2nd, the captain sent me on shore, to ascertain whether the chiefs had returned, and I was informed by the natives that they had, and were then at a house half a mile distant. This intelligence having been communicated to the captain, he went on shore, and took myself and Hussey for interpreters; but we found on our arrival, that the natives had been practising a piece of deception—the chiefs not having returned. Very much displeased at this perfidious treatment, the captain made a demand of the chiefs before sunset, threatening, if it were not complied with, to go on shore with fifty men, well armed, and destroy every person he could find.

This threat threw the natives into consternation, and immediately another messenger was despatched for the chiefs. The natives were so alarmed, that they soon sent off three or four more messengers; and we returned on board to dine. After dinner, I went on shore with Mr. Paulding, the first lieutenant, and some of the under officers, for the purpose of shooting birds. After rambling round the island for some time, we discovered a number of natives quickly approaching us from the lower part of the island; and supposing the chiefs were with them, we sat down to await their arrival; but before they came to us, a signal was set on board the schooner, for us to return, which was immediately obeyed, without waiting for an interview with the natives.

Early on the next morning, I was sent ashore to ascertain whether the chiefs had arrived, and soon found that they had, and were in a hut, waiting to receive a visit from the captain, who, I informed them, would come on shore after breakfast, to have a talk with them, and also to bestow some presents. Accordingly, the captain, with myself and Hussey, repaired to the hut, where we found them sitting, and ready to commune with us.

The captain told them he had been sent out by the Head Chief of his country, to look for the men that had been left there by the ship *Globe*—that he had been informed they murdered all but two—that, as it was their first offence of the kind, their ignorance would plead an excuse—but if they should ever kill or injure another white man, who was from any vessel or wreck, or who might be left among them, our country would send a naval force, and exterminate every soul on the island; and also destroy their fruit trees, provisions, &c. and that if they would always treat white men kindly, they never would receive any injury from them, but would have their kindness and hospitality reciprocated.

He also adverted to the practice of stealing, lying, and other immoralities; stating to the natives that these crimes are abhorred and punished in our country; and that murder is punished with death. He then sent me to the boat, lying at the beach, to bring three tomahawks, one axe, a bag of beads, and a number of cotton handkerchiefs, which were presented to the chiefs. He also gave them two hogs, and a couple of cats, with injunctions not to destroy them, that they might multiply. The captain caused potatoes, corn, pumpkins, and many valuable seeds to be planted, and gave the natives instructions how to raise and preserve them. He then explained to them that these acts of kindness and generosity were extended, because they saved us alive, and had taken care of us while among them.

This conversation with the natives being ended, we went on board, dined, and the captain and Hussey went again on shore. The first lieutenant made preparations for cruising in the launch, round the island, to make topographical surveys, who took me with him, as interpreter, and about 4 o'clock, we commenced a cruise with a design to sail up an inlet or inland sea; but the wind blowing fresh, and having a head sea, at 12 o'clock we anchored for the night.

Dec. 4th. At sunrise, we found ourselves not more than a mile from the place where we crossed over the evening before; and immediately getting under weigh, and rowing to the westward, we soon came to the place where the *Globe*'s station had been; anchored, and went on shore, for the purpose of disinterring the bones of Comstock, who had been buried there, and to obtain a cutlass, which was buried with him; but before we had accomplished the undertaking, the schooner got under weigh, and soon anchored abreast of us, at the same place where the *Globe*'s provisions were landed.

The captain and Hussey immediately came on shore to view the place; but as I caught cold the preceding night, by lying exposed in our launch, I was excused from serving further with Mr. Paulding in making surveys, and Hussey supplied my place. Soon after, I went on board with the captain, carrying with me the skull of the person we had dug up, and the cutlass, intending to convey them to America.

After dinner, the captain made a trip in the gig, to Alloo, taking me for his interpreter, where we arrived in half an hour, and soon travelled up to the village. The natives received us with marks of gladness, and in a short time the house at which we stopped was surrounded by them, who came undoubtedly for the purpose of gratifying their curiosity, by gazing at us. We remained at the village about two hours, during which time we had considerable talk with two of the chief women, and made some small presents to the people, such as beads, &c. They did not treat us as they usually do visitors, with fruit, &c. there being at that time what we call a famine, which in their language, is *Ingathah*.

After having taken leave of the natives, and walked about half the distance to the shore, we stopped to refresh ourselves under a fine cool shade. While in conversation on the manners and customs of the natives, an old man and woman approached us, who had acted towards me, during my residence among them, as father and mother. I immediately made them and their kindness to me known to the captain, who, in consideration of their humane treatment, rewarded them with a few beads and a handkerchief, for which they appeared thankful and grateful—telling them at the same time, the presents were to recompense their hospitality to me, and enjoining on them at all times to be friendly to the whites, and a reward would certainly await them. It being near the close of the day, we left Alloo, and having a fair wind, reached the schooner before dark.

The next morning, Dec. 5th, being very pleasant, all hands were employed in procuring wood for the schooner—some in cutting it down, and others in boating it off. Our carpenter had been engaged for a few days, at Milly; to instruct and assist the natives in repairing a canoe. The distance was four or five miles, and the captain wanting the carpenter, set sail for Milly in his gig, and soon arrived there; where he learned that the carpenter had repaired the canoe, to the great satisfaction of the natives, who expressed a strong desire that he might be permitted to remain among them on the island; but the captain informed them he could not spare him.

When the natives saw the carpenter packing up his tools, they expressed to me an expectation that the tools would be left with them as a present. We left the natives, and reached the schooner a little before sunset; the captain feeling anxious for the fate of the launch, as nothing yet had been heard of the fortune which had attended her, or the men in her.

Dec. 6th. Having procured a sufficient supply of wood, though our supply of provisions was hardly sufficient for the voyage, and the launch having returned, at about 10 a. m. we weighed anchor and proceeded to the place called Milly, where we anchored for the purpose of planting some seeds, and taking a last farewell of the chiefs and their people. The captain went immediately on shore, taking Hussey for his interpreter. He was gone till nearly night, when he returned, bringing with him Luttnon and several other natives. The captain gave orders to beat to quarters, to exhibit the men to the natives, and explain to them the manner of our fighting.

Those untutored children of nature, seemed highly gratified with the manoeuvres, but were most delighted with the music, probably the first of the kind they ever heard. We informed them we always have such music when we are fighting an enemy. The natives were then landed, and we immediately made sail for the head of the island, intending to cruise around the other shores of it, for the purpose of making surveys, and constructing a map of it. We stood eastward till nearly morning, then altered our course and headed towards the island.

During the following day, Dec. 7th, having favourable winds and weather, we made a regular survey of the whole length of the group, before sunset.—The captain now steered N. W. to endeavour to discover other islands which the natives had often described to me, during my abode with them. They said they had frequently visited ten or twelve different islands in their canoes, and that the people who inhabit them, all speak the same language, which is the same as their own, and that the islands lie about one day's sail from each other.

Dec. 8. The weather pleasant and fair; about 9 o'clock, a. m. we saw land ahead, and passed it on the windward side, then varied our course and sailed to the leeward of the island; but night coming on, we were obliged to defer landing till morning. The captain then attempted to reach the shore in the gig, but was not able to land, on account of the surf. After he returned on board, we made sail, cruising farther to the leeward, in hopes of finding a place to anchor, but in this we were disappointed, not being able to find bottom thirty yards from the rocks.

However, at high water, the captain, at imminent hazard in passing the surf, succeeded in landing. He had previously given orders to me and Hussey, not to let the natives know that we could converse with, or understand them, but to be attentive to everything that might pass among them, to ascertain whether their intentions and dispositions were hostile or friendly. After landing, the captain and Hussey visited the house where the head chief, or king of all those islands lived, of whom I had formerly heard so much, while I was on the Mulgraves.—They continued with him about two hours, were treated well, and discovering nothing unfriendly in the natives, the captain told Hussey he might make them acquainted with his knowledge of their language, by conversing with them.

The king, on hearing Hussey speaking in the language of the natives, appeared at first so frightened and agitated, that he could scarcely reply; but by degrees became composed, and inquired of Hussey where he learned their language, and why he had not spoken to them immediately on coming ashore. Hussey then informed him he was one of the two persons that had been on the Mulgraves, (in their language, Milly,) and that the other person (myself) was on board the schooner—that the schooner had been there after us, that we left the Mulgraves the day before, and had then visited that island for the purpose of examining it, &c. &c.

The king had long before heard of our being at the Mulgraves, and told Hussey he had been repairing his canoe, in order to go to those islands, with a view to induce us to live with him, who, had that been the case, would undoubtedly have used us well. The king was about seventy years of age, and had a daughter on the island where we had resided, wife to Luttnon. He inquired if his daughter was alive and well, with tears in his eyes and trembling form, for it was a long time since he had received any intelligence of her; and hearing of her welfare so unexpectedly, quite overcame the good old father's feelings. And here the reader will observe, that the pure and unaffected emotions produced by parental affection, are similar among all the human species, whether civilized or savage.

The natives of the island we were then visiting, may be ranked with those that have made the fewest approaches towards the refined improvements of enlightened nations, yet the ground work of humanity was discovered to be the same; and the solicitude of a fond father for a beloved child, was manifested in a manner which would not disgrace those who move in the most elevated circles of civilized life.

The old king expressed his regret that he had not visited the Mulgraves during our stay there, was very sorry we were about to return to America, and used all the force of native eloquence, to persuade us to continue with him. He inquired if we had got the whale boat he had heard of our having at the Mulgraves. Hussey informed him it was on board the schooner, and the swivel likewise. The captain then informed the king that he wanted cocoanuts and *bup*, which were obtained; and in return, the captain gave the natives some beads and handkerchiefs. The captain then went on board the schooner, made sail, standing a N.W. course, in pursuit of another island.

Dec 9th. About 10 o'clock in the forenoon, we discovered land ahead and off our lee bow. About 2 o'clock, p. m. we arrived near the land, hove the schooner to, and sent two boats ashore, to get provisions. At sunset the boats returned, loaded with cocoanuts and *bup*. We hoisted up our boats, and with a strong breeze, it being the inclement season of the year, prosecuted our voyage to the Sandwich Islands, & had much boisterous weather during the passage.

On Jan. 8th, 1826, we expected to make one of the Sandwich Islands, called Bird's Island, but night came on before we discovered it. But early on the following morning, we saw land about four leagues to the leeward, and bore down to the island for the purpose of sending a boat ashore, to kill seals.—We arrived near the landing place, hove to, and the captain with six men went ashore in the whale boat. We now stood off from the shore for about an hour, then tacked and stood in, for the boat to come off.

The wind had increased to almost a gale, and continuing to blow harder, when we were within a quarter of a mile of the island, not discovering anything of the boat, we veered off again, and continued tacking till night came on, but saw nothing of the boat or her crew. About 9 or 10 o'clock, the wind abated, and we found ourselves two leagues to the leeward of the island, where we lay to all night under easy sail, anxiously waiting for the approach of morning, in hopes then to learn the fate of the captain and men who had gone on shore.

At length the horizon was lighted by the dawn of day, which was succeeded by the opening of a very pleasant morning. We immediately made all sail for the island, but having a head wind, we did not arrive at the landing till near the middle of the day. A boat was sent on shore to learn what had befallen the crew of the whale boat, and shortly returned with all the men except the captain and one man that could not swim. We ascertained, that in attempting to come off through the surf, they were swamped and lost their boat.

We a second time sent the boat ashore with means to get the captain and other man, who were soon brought on board. We now made sail and steered our course for Woahoo, one of the Sandwich Islands, and nothing very material occurring on our passage, we anchored in the harbour of that island on the 14th. On the 16th procured a supply of fresh provisions. On the 19th, Hussey and myself went on shore for the purpose of rambling round the island, but nothing occurred worthy of notice.

Our foremast being found rotten a few feet below the top, it was deemed necessary to take it out for repairs, which required the daily employment of the carpenter and others for some time.—On the 27th, the captain received a letter, giving intelligence that the ship *London* had been driven ashore at an island not far distant from Woahoo.—As the *Dolphin's* foremast was out, the captain was under the necessity of pressing the brig *Convoy*, of Boston, and putting on board of her about ninety of his own men, taking with him 2 of his lieutenants and some under officers, he sailed to the assistance of the ship *London*.

Feb. 3rd, the brig *Convoy* returned laden with a part of the cargo of the *London*, and the *specie* which was in her at the time of her going ashore, under the command of our 2nd lieutenant, leaving the remainder of her cargo in another vessel, under the command of Captain Percival.

Feb. 5th. The captain returned with the residue of the *London's* cargo, and the officers and crew of that ship. After the cargo of the *London* had been secured, we were employed in finishing the repairs on our foremast, which were completed on the 21st; and we commenced rigging.

Feb. 26th. On the morning of this day, permission was granted to a number of our crew, to go on shore. In the afternoon, Hussey and myself went and took a walk. About 4 or 5 o'clock, I observed a great collection of natives, and on inquiring the reason, learned that several of the *Dolphin's* crew, joined by some from other ships lying in port, had made an assault upon Mr. Bingham, the missionary, in consequence of ill will towards that gentleman, strongly felt by some of the sailors, but for what particular reason, I did not distinctly ascertain. They carried their revenge so far, that they not only inflicted blows upon Mr. Bingham, but attacked the house of a chief. The natives, some with cutlasses, and others with guns, repelled the unjustifiable attack; and during the affray, several of our men were slightly injured, and one badly wounded, whose life was despaired of for some time. The offenders were arrested, sent on board, and put in irons.

On the next day, 27th, Mr. Bingham came on board with the captain and witnesses against the men engaged the preceding day, in the assault on shore. After a fair examination of evidence in the case, the aggressors were properly punished, and ordered to their duty.—The whale ships now began to arrive for the purpose of recruiting, and for some particular reasons, several of the captains of those ships requested Captain Percival to remain at the island as a protection to them, till they could obtain the necessary supplies, and resume their cruises.

From the present date, nothing of importance occurred that would be interesting to readers, till April 3rd, when great preparations were made on board the *Dolphin*, to give a splendid entertainment to the young king. The gig and second cutter were employed in the morning, to borrow signals from the different ships in the harbour, in order to dress out the schooner in a fanciful style. About 11 o'clock, the gig and second cutter were sent ashore for the king and several chiefs and natives of distinction, who were soon conveyed on board. The yards were manned, and a general salute fired. After partaking of as good a dinner as our resources and the means within our reach would afford, the king and his attendants were disembarked under the honour of another salute.—During the remainder of this month, the events which transpired, were principally of an ordinary cast, and not thought worthy of record.

May 3rd. This day we were employed in bending sails; and from this date to the 11th, the necessary preparations were made to commence our homeward voyage. This day (11th,) the pilot came on board, and for the last time we weighed our anchors in the harbour of Woahoo. While retiring from the shore we were saluted with 21 guns from the fort. We hove about, returned the salute, and then resumed our destined course, and bid a last *adieu* to Woahoo, after a tedious and protracted stay of about four months.

From the time of our departure, on the 11th of May, from Woahoo, nothing of importance transpired till the 12th of June. On the morning of this day we discovered the island Toobowy; and at 9 o'clock saw a sail, which proved to be a whale ship. At half past 2 came to anchor at a convenient place near the island, and sent a boat ashore, which returned at night with two natives, who gave us a description of the harbour, and directions how to enter it; and as our mainmast was injured, we entered it to make the necessary repairs.

On the 13th, we beat up the harbour, and at 3 o'clock anchored, where we continued repairing our mast, and procuring wood and water, till the 22nd; when we weighed anchor and made sail for Valparaiso, favoured with fine weather and good winds. July 18th, made the island of Massafuero, and passed it about midnight. On the 19th, in the forenoon, made the island of Juanfernandez; and at 11 p. m. on the following day, discovered the land at the south of Valparaiso.

On the 22nd, beat up the harbour, and at 2 o'clock on the morning of the 23rd, came to anchor.—At Valparaiso, we learned that the frigate *United States* was at Callao; and after getting a supply of provisions, we sailed for Callao on the 9th of August, and arrived on the 24th. Here we found the *United States*, lying under the island of Lorenzo, with several English ships of war.

On the 26th, the *Dolphin* in company with the *United States*, passed over to Callao; and Sept. 1st, I and the crew of the *Dolphin* were transferred to the *United States*.

Sept. 10th. All the men that had been transferred from the *Dolphin* to the *United States*, had liberty to go to Lima; at 12 o'clock we went on shore, and at 4 p. m. entered the gates of the city. I employed my time while on shore, in roving about the city, and viewing the various objects it presents; and on the 13th returned on board the *United States*. We were detained here till the 16th of December, when we sailed for Valparaiso, and having a pleasant passage, arrived on the 6th of January, where we were happy to find, for our relief, the *Brandywine*.

From the 8th to the 24th, all hands were engaged in preparing the ship for her homeward voyage; when at 9 o'clock we weighed our larboard anchor, and at 1 p. m. were under sail, passing out of the harbour, when the *Cambridge*, (an English 74,) then lying in the harbour, gave us three cheers, which we returned with three times three; she then saluted us with 13 guns, which we returned with the same number, and then proceeded to sea.

Being favoured with fine weather and good winds, we had a prosperous voyage to Cape Horn, and arrived off the pitch on the 7th of Feb. and passed round with a pleasant breeze. In prosecuting our voyage home, off the mouth of the river Rio de la Plata, and along the coast of Brazil, we had rough weather and thick fogs. On the 6th we made the land and harbour of St. Salvador, and about 9 o'clock came to anchor.—On the 7th we fired a salute for the fort, which was returned.

We were now employed in watering our ship, and making other

preparations for continuing our voyage homeward; and on the 15th got under weigh, with a fine breeze.

April 1st. At 10 o'clock, made the island of Barbadoes, and at 1 p.m. came to anchor, where we lay till 5 p. m. on the 3rd, when we got under weigh, and sailed down the island to St. Thomas, where we sent a boat ashore, and after transacting the business for which we stopped, made sail on the 9th for the port of New-York. On the 21st, made the highlands of Neversink; at 2 p. m. took a pilot on board, but owing to fogs and calms, did not arrive to the port of destination till 1 p. m. next day, when we anchored opposite the West Battery, with a thankful heart that I was once more within the United States.

CHAPTER 7

Observations of the Islands Visited

I will now proceed to give the reader some account of the islands I visited, and of the manners and customs of the natives, and shall endeavour to be as candid and correct as possible.

The Mulgrave islands are situated between 5° and 6° north latitude, and between 170° and 174° of east longitude. They are about fifty miles in length, and lie in the form of a semi-circle, forming a kind of inland sea or lake; the distance across it being about twenty miles. The land is narrow, and the widest place is probably not more than half a mile.

On the north side of the group are several inlets or passages, of sufficient depth to admit the free navigation of the largest ships; and if explored, excellent harbours would in all probability be found. In the inland sea are numerous beds of coral, which appear to be constantly forming and increasing. These coral beds are seen at low water, but are all overflowed at high tide. The whole group is entirely destitute of mountains, and even hills, the highest land not being more than six feet above the level of the sea at high water.

By the accounts given me from the natives, it appears that some parts have been overflowed by the sea. Their being so low, makes the navigation near them very dangerous in the night, both because they would not be easily seen, and because the water is very deep quite to the shores; and a place for anchoring can scarcely be found on the outside of the island.

The air of these islands is pure, and the climate hot; but the heat is rendered less oppressive by the trade winds, which blow constantly, and keep the atmosphere healthful and salubrious for so low a latitude.

The soil, in general, is productive of little besides trees and shrubs, and most of it is covered with rough coral stones.

The productions are breadfruit in its proper season, and cocoanuts, which they have throughout the year; and a kind of fruit different from any that grows in America, which the natives call *bup*—all growing spontaneously. Of the leaves of the trees the women manufacture very elegant mats, which they wear as blankets and clothing; of the bark of a vine they make men's clothing; and of the husks of the cocoa they make ropes and rigging for their canoes, and for almost every other purpose. The waters round the islands abound with fish, and the natives are very expert in catching them.

There are no animals on the islands, excepting rats; and by these little quadrupeds they are literally overrun.

The number of all the inhabitants, men, women, and children, is probably between five and six hundred.

The following may be given as prominent characteristics of the natives.—They are in general, well made and handsome—very indolent and superstitious. They are morose, treacherous, ferociously passionate, and unfriendly to all other natives. When they are not fishing, or otherwise employed, they are generally travelling about, and visiting each other. They have no salutations when they meet, but sit down without exchanging a word of civility for some minutes; but after a silent pause, the head of the family, if there is anything in the house to eat, presents it to his guests, who, when they have eaten sufficiently, if there are any fragments left, are very careful to secure them and carry them off when they return home; and the host would regard it as an imposition, if his visitors were to neglect this important trait of politeness, and fashionable item in etiquette.

They accustom themselves to frequent bathing; and commence with their children on the day of their birth, and continue the practice twice a day, regularly, till they are two years old. They do this to invigorate the system, and render the skin of their children thick and tough by exposure. Their living consists simply of breadfruit, cocoanuts, and *bup*; but cocoanuts are all they can depend on the year round—the two other articles being common only a part of the year.

Their diversions consist in singing, dancing, and beating time with their arms, in a manner similar to the amusements of the natives at the Sandwich Islands; in which they appear to take great delight.

They wear their hair long, and tie it up in a kind of bow on the top of the head, and this is all the covering they have for their heads. The men have long beards. One part of their dress makes a singular and ludicrous appearance, which resembles two horse tails suspended from the waist, one before and the other behind. The women's dress consists of two mats, about the size of a small pocket-handkerchief, which they tie round them like an apron.

I never saw any form of marriage among them, but when a couple are desirous of being united, their parents have a talk together on the subject, and if the parties all agree to the union, the couple commence living together as man and wife; and I never knew of an instance of separation between them after they had any family. In a few instances polygamy prevailed.

The following will give a pretty correct idea of their funeral rites and solemnities:

When a person dies, the inhabitants of the village assemble together, and commence drumming and singing, halloing and yelling; and continue their boisterous lamentations for about 48 hours, day and night, relieving each other as they require. This they do, because they imagine it is diverting to the person deceased. They bury the body at a particular place back of their houses, and use mats for a coffin.

After the ceremony of interment is performed, they plant two cocoanut trees, one at the head and the other at the feet of the buried person. But if the trees ever bear fruit, the women are prohibited from eating thereof, for fear of displeasing the bad spirit, *Anit*. And here it may not be inappropriate to remind the reader that Eve ate of the forbidden fruit, notwithstanding she knew it would displease the GOOD SPIRIT.

In their personal appearance, the natives are about the middle size, with broad faces, flat noses, black hair and eyes, and large mouths.

In relation to literature, they are as ignorant as it is possible for people to be, having not the most distant idea of letters.

Concerning the religion of the untaught natives of the Mulgraves, the following remarks will give all the knowledge I am in possession of:

They believe there is an invisible spirit that rules and governs all events, and that he is the cause of all their sickness and distress;—consequently they consider him to be a very bad being.—But they have no belief in a good spirit, nor have they any modes of worship.—It is a prevalent opinion among them, when any are sick, that the bad spirit rests upon them; and they believe that particular manoeuvres and a form of words, performed round and said over the sick, will induce *Anit*, the bad spirit, to cease from afflicting, and leave the unfortunate sufferers.

With regard to a future state of existence, they believe that the shadow, or what survives the body, is, after death, entirely happy; that it roves about at pleasure, and takes much delight in beholding everything that is transacted in this world;—and as they consider the world as an extensive plain, they suppose the disembodied spirits travel quite to the edge of the skies, where they think white people live, and then back again to their native Isles; and at times they fancy they can hear the spirits of departed friends whistling round their houses, and noticing all the transactions of the living. Singular as some of these notions and opinions may appear, there is much to be met with in Christendom equally at variance with reason; and I have heard from the pulpit, in New-England, the following language:

> I have no doubt in my own mind that the blessed in Heaven look down on all the friends and scenes they left behind, and are fully sensible of all things that take place on earth!

CHAPTER 8

Cyrus Hussey's Journal

This chapter, and the concluding remarks of the narrative, will be collated from a *Journal* kept by Cyrus M. Hussey; and if there appear occasionally some incidents similar to those recorded in the preceding account, it is believed the value and interest of this history will not be diminished by them.—Hussey commences thus:

About the last of April, myself and Lay were separated, destined to different islands, not knowing whether we should ever see each other again. At night we arrived at an island, and hauled up our canoe. We found but few natives, but among the number was the mother of the chief with whom I lived. She was very inquisitive respecting me, and talked so incessantly through the night that I could not sleep. The next morning we were employed in gathering breadfruit, for the purpose of curing it for the winter. This employment continued about three months, during which time I was very uneasy about my situation.

At intervals of leisure, when the old chief had no particular engagements to engross his attention, he would launch his canoe and go and search for fish; but my shoes having been taken from me, whenever I was employed round the rough shores of the island, my feet were so wounded that I could hardly walk. The natives now commenced the destruction of my clothing, and not being able to converse with them, I found it very difficult to preserve my apparel. They often requested me to divest myself of my clothing, and dress as they did, or rather not dress at all. I made signs that the sun would burn me, if I should expose myself to its scorching rays.

When they found that persuasion would not induce me to divest myself of clothing, they began to destroy my clothes, by tearing them in pieces. It was some time before I could understand their language, so as to inform them that the sun would burn my back; and being robbed of my clothes, the powerful influence of the sun soon scorched me to such a degree that I could scarcely lie down or take any rest.

About the latter part of July, William Lay and others came to the island in a canoe, to see me, being the first interview we had enjoyed since our separation, which was about three months previous. Lay informed me that the natives had taken his bible from him and torn it up, and threatened his life. He informed me that it seemed to him as though he was robbed of that comfort which none in a Christian land are deprived of. We were soon parted; he in a canoe was taken to an island by the natives called Dilabu, and I went to my employment, repairing a canoe which was on the stocks.

After I had finished the canoe, the natives prepared a quantity of breadfruit and fish for the chiefs, and on the following morning we set sail for an island called Milly, one of the largest in the group, at which resides the principal chief. We arrived just at night and were cordially received by the natives, who had assembled on the beach in great numbers, for the purpose of getting some fish which the old chief had brought with him. He then hauled his canoe on shore; and I had again the pleasure of seeing my fellow sufferer, William Lay, after a month's separation. Since our first meeting we were not allowed to converse much together.

The old chief tarried at this island but a short time, and Lay and myself were once more separated. The old chief, his family, and myself, returned to the island which we had left two or three days before, called, in the language of the natives, Tabarawort; and he and his family commenced gathering breadfruit. As the old man with whom I lived had charge of several small islands, we found it difficult to gather the fruit as fast as it ripened, so that a considerable part fell to the ground and perished. In the meantime, while we were employed in gathering in the fruits of the earth, news came to the island, to inform the chief with whom I lived, that it was the intention of the highest chiefs to destroy us both, (that is myself and Lay,) because a severe sickness prevailed among them, and they being superstitious, supposed we were the occasion of it.

I informed them that we could not have been the cause of the sickness, as no such sickness prevailed in our country, and that I never before had seen a similar disease. But still they talked very hard about us; and the highest chief sent to the chief I lived with, to have me brought to the island of Milly, where Lay lived, in order that we might be killed together. Preparations having been made, the old chief, whom I called father, with his family and myself, set sail the next morning for Milly, where we arrived about sun set. He immediately went to see the chief of Milly, to inquire the circumstances relating to the necessity of taking our lives, leaving me and the rest of the family in the canoe.

I shortly perceived William Lay and his master coming towards the canoe, which produced sensations hard to be described. Affectionate and sympathizing reader, what must have been our feelings and conversation at that moment, when nothing seemingly was presented to our view but death? We were allowed an interview of only a few minutes, when we were again separated.

My master soon returned to the canoe, and entered into very earnest conversation with his family, which, at the time, I did not fully understand; but found afterwards it was a relation to his family of his interview with the natives on the subject of taking our lives; and that if they killed me, they would first have to kill him, (my master,) which they were unwilling to do. My kind old master told them he had preserved me, and always should. Night now coming on, I lay down to sleep, but fear had taken such possession of my mind, that the night was spent in wakeful anxiety.

The next morning I asked leave of my master to visit Lay, which he readily gave. I set out for the hut in company with my master's son; but on approaching it, Lay called out to me, to inform me that I must not come—that the natives did not like to have us together. On my turning to go back, Lay's master called to me to come. I went and sat down, and entered into conversation with Lay, to ascertain what the intention of the natives towards us were. He told me it was the design of the high chief to kill us. I observed to him, that we were in the hands of the natives; still there was a higher and more powerful Hand that could protect us, if it were the Divine pleasure so to do. I then bade him farewell, and returned to the canoe, never expecting to see each other again till we should meet on the tranquil ocean of eternity.

My master being now ready to return to his island, the canoe was launched, and we set sail, and arrived the same night, having been absent two days.—The natives expressed much joy on seeing me return, and asked many questions respecting the chief of Milly; but as I was unable to speak their language intelligibly, I could give them but little information. We then went on with our work as usual, which was fishing, &c. &c.

After having been at this island some time, my master's wife manifested an inclination to go and visit her friends, who lived at an island called in their language Luguonewort. After a successful excursion in fishing, we cooked a part, and took some breadfruit, and embarked, agreeably to the wishes of my master's wife, and arrived at Luguonewort in two days. The natives of that island gave us a cordial reception. We hauled up our canoe and remained some time among them. After our agreeable visit was ended, we returned to the other island, found the natives well, and that good care had been taken by the chief's mother, an old woman to whom the superintendence of things had been left.

About six months after the massacre of my shipmates, the brother of the native in whose possession I was, came to the island, and informed us that a ship had been seen to pass a day or two before, and that it caused great disturbance among the chiefs—that they thought it was the ship that left the islands, (the *Globe*,) and that she was in search of us. My old master immediately prepared his canoe to visit the chiefs, and he wanted also to inquire of me what I thought respecting the ship.

We loaded our canoe and made sail for Milly, where the chiefs were. We arrived at night, and found a great number of natives collected on the beach, to see if we had any fish. We hauled up our canoe for the night, and the natives began to question me about the ship.—I told them I did not know, concluding it would be good policy to say but little on the subject.

The natives crowded round me in great numbers; and I did not see Lay till he came to me. I inquired of him what he had seen, and he informed me that there had been a ship in sight about half an hour before sunset, and that she was near enough for him to see them take in their fore and mizen top gallant sails, but could give no definite account of her, as she was soon out of sight.

We were not allowed to be together long; and I went to rest as usual, but could not sleep.—*Hope springs eternal in the human breast*—and hope that the ship which had been seen had come to deliver us from savages and transport us to our native country and dear friends, had an influence on my feelings more powerful than sleep, and imagination was busy through the night in picturing scenes of future happiness.

But the prospect of our being released from our unpleasant situation was not very flattering. Early next morning I asked and obtained permission from my master, to pay a visit to Lay, before passing round to the opposite side of the island. Accompanied by my master's son and several others, I went to the hut where Lay lived, and we had the pleasure of another interview; but it was of short duration, for we were not allowed to be together more than a quarter of an hour.

I returned to my master's canoe, and there continued till the middle of the day; we then launched and set sail for Tabanawort, where we arrived the fore part of the night.—Early next morning we prepared for a fishing cruise, had pretty good success, and returned just before night, made a fire, cooked some fish, and ate a delicious supper.

Our canoe being leaky and very much out of repair, my master and I commenced taking her to pieces, for the purpose of rebuilding her; and we were occasionally employed upon her nearly two months, when we launched her, and commencing fishing business, had alternately good and bad success. One day we had the good fortune to enclose, in a kind of weir made for the purpose, a large quantity of fishes, and with a scoop-net we caught a plentiful supply.

After cooking them, we set out with a quantity to dispose of to the chiefs of Milly, where we arrived before night, on the same day of sailing. Very soon after our arrival I saw Lay and his master approaching the canoe, and we once more had a short but pleasant interview. I inquired of Lay how he fared, as to food, &c. His reply was, better than he expected, and that the natives were kind to him, always giving him his part. I informed him I had a basket of fish reserved for him as a present, which he requested me to keep till dark, that he might be enabled to carry them home without having them all begged by the natives. He came at night for the fish, and I retired, agreeably to my master's wishes to sleep in the canoe, to prevent the natives from stealing the remainder of the fish that were on board.

The next morning my master was highly pleased to find that nothing was missing; and gave me liberty to go and see Lay. I went to the hut and found him with his master. They gave me a cordial welcome, and presented me with some cocoanuts in return for the fish. Lay's master inquired of me very particularly respecting my master, and the quantity of fish we caught. I then returned to the canoe, carrying the cocoanuts, to deposit in the hold.

My master asked me where I got them; I told him Lay's master gave them to me. If this minute detail should appear unimportant to the reader, he may draw a moral from it; for it evinces that my master was like other masters, desirous to know if his servant came honestly in possession of the cocoanuts. He then ordered me and his son to launch the canoe, which we did, got under sail for the island we left the day before, and arrived back at night. We learned that during our absence the natives had caught a considerable quantity of fish; and in a few days we caught a large quantity more; loaded our canoe, and embarked for one of the head islands to pay a visit, where we stopped some time.

On our return, we commenced catching a kind of fish called by the natives *kierick*. They are about the size of a small codfish; and the manner of taking them is very curious—they make a line of the husk of cocoanuts, about the size of a cod line; they then in the canoe pass round the fish to the windward of the flat, then lie to till a considerable quantity of them get on the flat, then square away by the wind and run down and go round the flat with this line, and thus catch them, men, women, and children being employed. I have known them catch one hundred at a draught.

The fish are afraid of the line, and when enclosed, taken by a scoop-net. After taking a sufficient quantity, they go on shore to prepare for cooking them, which is done by digging a large hole in the earth, filling it with wood, covered with stones. The wood is then consumed, which heats the stones—the fish are wrapped in leaves to prevent them from falling to pieces, then covered with green leaves, and cooked by the heat of the stones. About an hour is required to cook them sufficient for eating. Their manner of curing fish, is, to split them and dry them in the sun, without using salt. Thus cured, they will keep some time.

While we were employed in fishing, Lay came to the island, in company with a native, to visit me; but did not stay long, for the chief sent for him, fearing, as I afterwards found out, that they should lose us. From some hints that had been dropped, a report had got in circulation that my master and Lamawoot, (Lay's master,) intended to leave their islands, and embark for an island to the north west, where the king lived, and carry us with them as a great curiosity.

Lay was carried back to the chiefs—the head one sent a command to my master and Lay's to come and see him—they made preparations and set sail for Milly; where they were closely questioned respecting their going to the other island, &c. &c. They denied that they had even intimated any such design; which was false, for I had frequently heard them talking on the subject myself, but kept silent, as it appeared to be a great crime for any to desert their islands; and I feared the consequences of making it known.—They then parted in peace and friendship, and I and my master returned to our habitation.

We then went to an island to catch fish, and a disagreement taking place between two of the natives, about some trifling affair, the particulars of which I did not learn, one of them took a spear belonging to the other, and after breaking it across his knee, with one half of it killed his antagonist, and left him. The parents of the man killed, being present, laid him out on some mats, and appeared to regret their loss very much. They kept a continual drumming over the body of the deceased for two or three days; after which he received a decent burial on another island at some distance from the island where he was killed.

Chapter 9

Journal Continued

Having a successful fishing voyage, we loaded our canoe, and carried our cargo to the chiefs of Luguonewort. I had the satisfaction of an interview with Lay; but our provisions being soon exhausted, we were obliged to go again in search of fish. At this time there was a severe drought, and breadfruit trees suffered extremely, many of them entirely died. The superstitious natives supposed the drought was sent upon them as a judgment, because myself and Lay were allowed to live. I informed them that we could neither make it rain nor prevent it; but some of them were so ignorant that they believed we could control the weather.

But some of the chiefs thought the drought was visited upon them because they had killed our shipmates, and I was always ready to join with them in that opinion. The drought continued about four months with such severity that most of the breadfruit trees on the small islands were so completely dried up that they never sprouted again. Many of the ignorant natives still insisted that their sickness and drought were occasioned by suffering us to live upon their islands; but this gross ignorance was counterbalanced by most of the chiefs, who believed differently, and to their more liberal opinion we are indebted for our lives.

About this time the islands were refreshed by plentiful showers of rain, and the natives assembled at Milly to sing for the breadfruit to come in abundance. They said their singing would please *Anit*, and that he would reward them with a very great crop.

A disturbance existed between the high chief and his brother Longerene. The disagreement lasted about nine months, during which time the two brothers did not see or speak to each other. Luttuon, the high chief, then sent a canoe to inform his brother Longerene that he wished to see him. An interview took place, and a treaty of peace was ratified.

During our stay at Milly, I had frequent opportunities of seeing Lay, my fellow sufferer; but the only relief we could afford each other was derived from a sympathy of feelings, and in conversations relating to our homes and native country, by blending our mutual wishes for a safe return, &c. &c. The reader can hardly conceive the unpleasantness of our situation at this time—the famine was so great that the tender branches of trees were cooked, and the nutritious juice drank as food.

My strength was so reduced in consequence of being deprived of my usual quantity of provisions, that I was unable to accompany my master on a fishing voyage. When my master returned, he found me lying in the hut, and asked me what was the matter. I informed him my indisposition proceeded from hunger; he cooked a fish and gave me, which, though it afforded me some relief, was not half enough to satisfy the cravings of appetite.

After I had recruited my strength, one day while engaged in fishing, a canoe came to the island; and as soon as the canoe was near enough for the natives in her to be heard, they commenced hallooing and making dreadful noises, which is their practice when war is declared. They informed us that the high chief had killed several of the lower chiefs who belonged to the island called Alloo; that Longerene had fled to Alloo, his own island; and that the high chief was determined to pursue and kill him. We were ordered to go immediately to his assistance; accordingly we set sail for the island Milly, where we found a great number of natives collected for war.

Again I had the satisfaction of being with Lay; who informed me that they were going to fight the other party at Alloo; and that the high chief had told him that he and I must prepare two muskets, and go and fight with them. Luttnon sent for me and Lay, and informed us he was about to have a battle, and that we must prepare to take a part in it. We asked him if he had any powder—he said he had a plenty, and showed us a small box, which contained a little powder and mustard seed mixed together, which, if it had been good powder, would not have made more than five or six charges. We told him it was good for nothing; but he said we must do the best we could with it.

As we were afraid to offend him, we went to work with the powder, and dried it in the sun, and prepared our muskets for battle.—The next morning we launched fifteen or sixteen canoes, containing in all about 200 natives, and set sail for Alloo; where we arrived and landed, and proceeded to a village in order to give battle to the enemy. On learning that the chief of Alloo and his family had fled in a canoe, we returned to our canoes, made sail in pursuit of the chief, but did not overtake him. After returning and spending a day or two at the island of Alloo, we launched our canoes and went to our respective homes, and heard no more of the war.

Sometime after my master returned to the island where we usually resided, a canoe came and brought the information that a vessel was anchored near one of the head islands—that she carried guns on each side, and had a hundred men—that they (the natives that brought the news) had been on board of the vessel, and received presents of beads, which they had on their necks. The natives said the vessel was not like our ship which we came in, but had only two masts. I told them we had vessels of all descriptions, some with one mast only.

They said the men on board did not look like us, and that they were very saucy. I informed the natives the vessel was a war vessel, and that if molested by the natives, they would shoot them. The natives said they would take the vessel and kill all the men on board. I told them their safety consisted in friendship, and that any hostile attack on the crew of the schooner would lead to their own destruction.— They then set sail for Milly, to inform the chiefs of the arrival of the vessel at the head island. The chiefs of Milly gave orders to launch the canoes, fifteen in number, to go and take the schooner. These canoes were manned by 200 natives. My master's canoe not being in perfect repair, we could not join the party.

On the night of the 25th, (Nov.) we saw several of the canoes returning towards the island where I was. From one of the canoes landed the high chief, who began to question me respecting the vessel. I told him I had not seen the vessel, and of course could not tell much about her; but that I expected she had come after me and Lay, and that she would have us. He then said he had better kill us both, and then there would be no one to tell that the natives had killed the rest of our crew. I told him that the people on board the schooner knew there were two alive, and if they killed us, the crew of the vessel would kill all the natives. This appeared to perplex his mind, and he shortly left me, and retired to rest.

On the next morning, 26th, the chief again questioned me respecting the vessel, but I could give him no particular information, as I had not seen her.—The natives then commenced knotting up leaves to inquire of their god, who, they said, would inform them what was best to be done. Towards night they departed, leaving me with my master, giving him strict orders not to let me go to the vessel, fearing that I should not only remain on board, but give information that my shipmates had been murdered. I was glad to see them depart, for I feared they would kill me.—The reader can have but a faint idea of my feelings at that time; nor will I attempt to describe them.

Towards the close of the next day, (27th,) a canoe came to the island which had been boarded by a boat from the schooner. The natives offered the men in the boat some cocoanuts, which they would not accept. The boat then proceeded towards the island of Milly.—The natives informed me that the men in the boat inquired after the men who were left there by the ship *Globe*; but they would not give any information where they were. The canoe left the island, and we went to rest.

The next day passed without hearing anything of the schooner; but the day following, (29th of Nov.) as I was walking in the woods in the afternoon, I heard a dreadful outcry for Hussey. I ran to the hut to learn the cause, and to my unspeakable joy, I discovered that one of the schooner's boats was on the beach, waiting for me, the men all armed and equipped for battle.

As I approached, the lieutenant spoke to me and told me to come to him. I went and sat down by him. He asked me several questions, but my feelings were so overcome and agitated, that I know not whether I replied in English, or the language of the natives. While we were sitting together, the old man whom I had always called master, but who was now willing to be considered my servant, asked me if the white people were going to kill him. The lieutenant inquired of me to know the purport of the old man's question; I told him he was afraid of being killed. The lieutenant replied that he should not be hurt, if he behaved himself properly.

We then walked round the island, and I collected what few things I had, a musket, &c. and made preparations for our departure. My old master being unwilling to part with me, asked permission to go with me. I spoke to the lieutenant on the subject, and he readily consented. We then set sail, accompanied by my master and his son. We soon fell in with the 2nd Lieutenant, in another boat, who informed that all the survivors of the *Globe*'s crew were now rescued. The boats soon lost sight of each other, as night came on, and that in which I was arrived at the island about 9 o'clock in the evening. We landed, cooked supper, and anchored our boat at a little distance from the shore for the night.

The next morning, (30th,) we got under weigh, accompanied by the other boat, beat to the windward, for the outside passage, and then ran down to the schooner, and got along side at 9 o'clock. I will leave it for the reader, to picture my feelings on entering once more on board of an American vessel, after having been among unmerciful savages 22 months. We soon had some breakfast, after which my hair was cut, which was of two year's growth, and I was furnished with clothing, and remained on board till the next day.

From this date to the time of our arrival in the United States, all the important incidents and facts which transpired, will be found in the preceding pages, arranged from the journal kept by Lay.

After expressing my thanks to all who assisted to rescue us from savage bondage, and my gratitude to Heaven for a safe return to my friends and native land, I bid the reader a respectful farewell.

Memorandum

Willam Rotch

WILLIAM ROTCH

Contents

Prefatory Note	91
Memorandum	93
The Respectful Petition	126
Answer of the President	130
Copy of Thomas Jenkins's Complaint	133

Prefatory Note

Though a man of some prominence in his day and a member of a family still well known and honoured in Massachusetts, the writer of this *Memorandum* is unnoticed in most books of reference, and it seems fitting to preface his reminiscences with a brief account of his life.

William Rotch was a Nantucket Quaker, born on the island, October 15, 1734. He carried on a large whaling and shipping business in Nantucket, which was for many years the third largest port of New England. Though he was a man of peace, his fortunes were closely connected with the fortunes of war. While visiting London in 1773 he chartered three of his ships to the East India Company, and these ships—the *Dartmouth*, the *Beaver*, and the *Bedford*—brought the tea to Boston and furnished the scene of the "Boston Tea Party," one of the early outbreaks of the revolutionary spirit in New England.

When the Revolution actually came it made serious trouble for the peace-loving inhabitants of Nantucket, who did their best to preserve their neutrality and keep the whaling industry alive to supply the world with oil. How William Rotch met these difficulties he himself relates in his *Memorandum*. He was successful both during and after the war, and his ship *Bedford* was the first vessel to carry the American flag into a British port, when, on February 6, 1783, she reported at the London custom-house with a cargo of oil. There is a story that one of her crew, a hunchback, while on shore one day, was clapped on the back by a British sailor, who said, "Hello, Jack, what have you got here?" to which the Yankee replied, "Bunker Hill, and be d—d to you!"

It may be worthwhile to note that William Rotch's son Benjamin, who accompanied his father on his mission to England and France, related two interesting anecdotes which are not included in the *Memorandum*. One of these is to the effect that when Mr. Rotch had completed his arrangements for sailing to France, Lord Hawkesbury sent word to him desiring another interview, and that the Quaker's reply was as follows: "If Lord Hawkesbury wishes to meet William Rotch, he will find him on board the ship *Maria* until the hour when the ship takes her anchor." The other story is that, during the visit to the royal chapel in Paris, the king himself, who shared the prevailing curiosity to see the wealthy Quaker, was present incognito.

As related in the *Memorandum*, William Rotch in 1795 removed to New Bedford, which was afterwards for many years the leading whaling port of the world. Here he lived until his death on May 16, 1828. He was a man of the highest principles, much respected and loved wherever he was known.

Boston, April, 1916.

Memorandum

A friend of mine has repeatedly requested me to put on paper some of the occurrences of about twenty years of my life from 1775 to 1794 which he had heard me relate in conversation.

When the Revolutionary War begun in 1775 I saw clearly that the only line of conduct to be pursued by us, the inhabitants of the Island of Nantucket was to take no part in the contest, and to endeavour to give no occasion of offence to either of the contending powers.

A great portion of the inhabitants were of the Denomination of Friends, and a large number of the considerate of other Societies united in the opinion that our safety was in a state of neutrality as far as it could be obtained, though we had no doubt that suffering would be our lot, which we often experienced from both parties. Our situation was rendered more difficult by having a few restless spirits amongst us, who had nothing to lose, and who were often thwarting our pacific plan, and subjecting us to danger, not caring what confusion they brought upon us, if they could get something in the scramble.

My own trials begun soon after the war broke out. In the year 1764 I had taken the goods of a merchant in Boston, deceased insolvent, who was deeply indebted to me.

Among these were a number of muskets, some with, and others without bayonets. The straits of Belleisle opened a new field for the Whale Fishery, where wild fowl were abundant, and my guns met with a rapid sale. Whenever those with bayonets were chosen, I took that instrument from them. The purchaser would insist on having it, as an appendage belonging to the gun, and I as strenuously withheld it, and laid them all by. Many years afterwards I removed to another store, leaving much rubbish in the one I had left. Among the rubbish were these bayonets, neglected and forgotten; until the war commenced, when to my surprise they were brought into view by an application for them, made by a person from the Continent.

The time was now come to endeavour to support our testimony against war, or abandon it, as this very instrument was a severe test. I could not hesitate which to choose, and therefore denied the applicant. My reason for not furnishing them was demanded, to which I readily answered, "As this instrument is purposely made and used for the destruction of mankind, I can put no weapon into a man's hand to destroy another, that I cannot use myself in the same way"—The person left me much dissatisfied. Others came, and received the same denial. It made a great noise in the country, and my life was threatened. I would gladly have beaten them into "pruning hooks," but I took an early opportunity of throwing them into the sea.

A short time after I was called before a committee appointed by the court then held at Watertown near Boston, and questioned amongst other things respecting my bayonets.

I gave a full account of my proceedings, and closed it with saying, "I sunk them in the bottom of the sea, I did it from principle, I have ever been glad that I had done it, and if I am wrong I am to be pitied." The chairman of the committee Major Hawley (a worthy character) then addressed the committee, and said "I believe Mr. Rotch has given us a candid account, and every man has a right to act consistently with his religious principles, but I am sorry that we could not have the bayonets, for we want them very much."

The major was desirous of knowing more of our principles on which I informed him as far as he enquired.

One of the committee in a pert manner observed "then your principles are passive obedience and non-resistance." I replied "No my friend, our principles are active obedience, or passive suffering." I had passed this no small trial respecting my bayonets, but the clamour against me long continued.

From the year 1775 to the end of the war, we were in continual embarrassments—Our vessels captured by the English, and our small vessels and boats sent to the various parts of the Continent for provisions, denied, and sent back empty, under pretence that we supplied the British, which was without the least foundation. Prohibitory laws were often made in consequence of these unfounded reports. By this inhuman conduct we were sometimes in danger of being starved. One of these laws was founded on an information from Governor Trumbull of Connecticut, who had been imposed upon respecting our conduct in supplying the British.

I wrote to the governor on the subject, and laid our distress very home to him, assuring him at the same time that nothing of that kind had taken place. He was convinced of his error, and was ever after very kind in assisting us within his jurisdiction.

But there were so many petty officers, as Committees of Safety, Inspection, etc. in all parts, and too many of them chosen much upon the principle of Jeroboam's Priests, that we were sorely afflicted.

It was about the year 1778 when the current in the country was very strong against us at Nantucket, the vessels we sent after provisions, sent back empty, and great suffering for want of food was likely to take place, that the people who thought we ought to have joined in the war (not Friends) began to chide and murmur against *me*. They considered me the principal cause that we did not unite in the war (which I knew was measurably the case,) when we might have been plentifully supplied, but were now likely to starve, little considering that if we had taken a part, there was nothing but supernatural aid (which we had no reason to expect) that could have prevented our destruction. Though I had done everything in my power for our preservation, this murmuring of the people operated so severely upon my spirits, that I was once (a time never to be forgotten) on the point of asking of that Divine Being who gave me life, that he would take it from me, for my affliction seemed more than I could bear. But being restrained by that good hand, which had so often been my deliverer, after shedding a flood of tears, my mind was more easy, and my spirit revived.

In the year 1779 seven armed vessels and transports with soldiers from Newport came to us, the latter commanded by George Leonard, an American, as were his troops in general, having joined the English. They plundered us of much property, some from me, but a considerable amount from Thomas Jenkins. While they were plundering his store, I attempted to pass the guard they had set, being desirous to see Leonard, and intercede with him to desist. But the guard arrested my progress with the bayonet. After some time Timothy Folger succeeded in speaking to him, and advised him to go off, for the people would not bear it much longer. He took the hint, and retired much enraged.

We soon had information that Leonard &. Co. were preparing another and a more formidable expedition to visit us. The town was convened to consult what measures should be taken in this trying emergency, which resulted in sending Dr. Benjamin Tupper, Samuel

Counting House of William Rotch & Sons at the foot of Main Street in Nantucket Built in 1772

Starbuck and myself to Newport, to represent our case to the commanders of the navy and army. We arrived in the harbour of Newport, where Captain Dawson commanded the navy, and General Prescott the army.

But the American refugees had made interest with the general not to suffer us to land, and we were ordered by Dawson to depart. We interceded with him to let us stay a little longer, for we found the expedition was progressing rapidly, and unless we could arrest it, it would be in vain to proceed to New York. Dawson by request of General Prescott, under the influence of the refugees, ordered our immediate departure again. Dr. Tupper now for the first time went on board, and in his plain blunt way, after the usual ceremony of entry, addressed him in this manner—

> You order us to depart. We cannot be frightened away, nor *will* we depart. We know the extent of your authority. You may make a prize of our vessel, and imprison us—much better for us to be thus treated, than to be sent away. We came here for peace, and you ought to encourage everything of this kind, etc.

His reasons made such an impression on Captain Dawson, that he gave us liberty to stay as long as we pleased.—The refugee boat came several times to us, to get us [to go] off.—We insisted on going on shore, but they as often refused us. After this conversation with Dawson, the boat came again, and Dr. Tupper insisted that he would go on shore

They still denied him unless he intended to stay with them. As he was not always exact in his expressions, to answer his purpose he says "Well, I am going to stay," and almost forcibly got into their boat, and went on shore, being satisfied that if he could once see the general, he could in this respect destroy the influence of the refugees. He accordingly got liberty for Samuel Starbuck to come on shore, and the next day for me to follow. We found it necessary to be in friendship with the refugees, that if possible we might stop the current of their intended predatory visit.

I got on shore in the afternoon, and found that I must wait on General Prescot.—Knowing his brittle temper, and it being in the *afternoon* I almost dreaded to appear in his presence. However, let my treatment be what it would, I wished it over and accordingly went.

I was introduced to him by one of his aids—He received me very cordially, gave me his hand, and said "Mr. Rotch will you have some dinner?—I can give you good bread, though the Rebels say we have none." I thanked him saying I had dined—"Well, will you take a glass of wine?"

"I answered "I have no objection if thou canst put up with my plain way"—The glass was filled, with his own, and those of all the officers at table—as a stranger introduced, they all drank to me before I put the glass to my lips—I then observed to the general, "As I mentioned before if thou couldst put up with my plain way, I was willing to take wine with thee, but as we as a society disuse these ceremonies, I have always found it best to keep to my profession, let me be in what company I may. Therefore I hope my not making a like return will not be accepted as any mark of disrespect, for I assure thee it is not the case."

His answer was, "Oh, no, if a Quaker will but *be* a Quaker, it is all I want of him—But —— is no Quaker"—(naming one of our profession) and I was sorry for the cause of his remark.

After some conversation, I mentioned that I did not wish to intrude further on his time, and rose to retire—"Oh no," says he, "you must take coffee." I accordingly stopped. He was full of conversation respecting the siege of the Americans, and made it a very trifling thing. I then mentioned (the French Fleet being at that time before the town) that twelve capital ships being before the town we thought was much against them.

"To be sure," said he, "it is not very pleasant, but we do not mind them."

I then gladly got away. We applied to Major Winslow formerly of Plymouth to introduce us to Colonel Fanning who was the principal. When we mentioned our situation, that we were likely to be destroyed, the colonel was very high, saying we might join the English then—We observed that such a step would inevitably destroy us.- "Well, said he, I have been destroyed also"—(I believe he was Governor of North Carolina at the beginning of the War)—Major Winslow endeavoured to soften him by representing our peculiar situation, but there seemed little prospect of anything favourable when we parted. They had a Board of Refugees established. Colonel Fanning President, who would hear us when they met. We accordingly attended, and found Fanning very mild, and disposed to alleviate our sufferings—and as we proposed applying to the commanders in New York, we asked if they would put off their expedition, until they could know the result of our mission there.

Fanning thought this reasonable, and put the question to all the principals there.—They readily agreed until it came to Leonard, who very reluctantly gave his assent.

We then proceeded to New York, and applied to the commanders, Commodore Sir George Collier of the navy, and Sir Henry Clinton of the army. On representing our case to Sir George, he readily gave us an order, forbidding any British armed vessel to take anything out of our harbour. This was a great relief.

I then laid before him the state of our captured seamen, that all the exchange of prisoners at that time was partial, that as we made no prisoners, we had none to exchange, consequently ours remained in the prison ships until they mostly died. On his understanding the reasonableness of the request, he ordered that all our men should be released that were not taken in armed vessels (for such we had no right to apply) and that it should not be so in future as long as he had the command.

We also applied to Sir Henry Clinton through one of his aids. (Major André that fine young man who lost his life as a spy.) We could get no written order from him, but he intimated that he would direct that those in his department should not molest us, which no doubt he did, as they gave over their expedition, and we had a little quiet, until Sir George Collier's command was superseded by the arrival of Admiral Arbuthnot, and the shaving mills then came upon us.

Timothy Folger was then sent to New York, and he obtained a like order as that from Sir George Collier. Added to this, he got Permits for a few vessels, about fifteen, to whale on our coast, which were successful, but it was with great difficulty that distinction could be made between British and American armed vessels, as the latter would make prize of us if British permits were found.

I now come to the most trying scene in my experience during the war,—which was being with four others impeached for high treason by Thomas Jenkins, where there was no step between being clear, and death.

The laws of this state at that time made it high treason for any person to go to a British port without the consent of the court.

We were well assured that if we applied we should be refused, and if we did not apply to the British, we should from every appearance be destroyed by them. Under serious consideration I was willing to be joined to the two others before mentioned and proceed, as with our lives in our hands. This was made one of the great charges among

The Rotch (Roach) Fleet among a school of sperm whales off the coast of Hawaii

The ships are the Enterprise, William Roach, Pocahontas, and Houqua.

others in the impeachment, a copy of which will be annexed.

We were examined before a committee of the court on the impeachment, but knowing we were to appear again when the witnesses should attend, we made no defence, which we afterwards found was an error in judgement.

By this means the court thought us guilty, and were about making out an order to the Grand Jury, to find a bill against us and commit us to prison, which if it had taken place would have been in the severe winter of 1780.

But happily my much valued friend Walter Spooner, a member of the court, just then arrived, perceived the business before the court, and came to us for information. We told him we had reserved our defence for the second examination. He considered us in an error, and said we must send for Jenkins who resided at Lynn, and have another examination, and he would get the court to stay their proceeding until this should be done.

We accordingly all met before the committee, General Ward a worthy character in the chair.

It was put to me first to answer to the high charges. (When I rose he politely told me I need not rise—I thanked him, but my heart was so full that my tongue seemed incapable of utterance while sitting.) I answered to the charges in such a manner as fully to convince the committee of our innocence—When it was finished, the chairman, I have no doubt from a desire to put our minds at ease, asked me when we expected to return home—I replied that he could better judge of that than I could—(being now in custody)—He then asked me if I would take the *subpoenas* for the witnesses to Nantucket, and deliver them to the sheriff. This was also to console our minds. I answered in the affirmative if he thought proper to entrust me with them. I accordingly took and delivered them.

In the spring following we appeared again with twelve out of twenty witnesses, who were all I could get to attend, and then had another full examination.

Before we entered upon it I desired liberty to ask Jenkins a question, which was granted. Some of his friends had propagated a report, that I had offered him money, if he would withdraw his complaint. I then put the question to him, whether I had ever made him any offer of the kind—but it irritated him—I therefore went no further in a question to him, but desired liberty to make my declaration, before we entered on the charges in the complaint, which was readily granted.

I then said, "I now declare in the most solemn manner, that I never have, directly or indirectly, by myself, or by any person for me, proposed or offered one farthing to Thomas Jenkins to withdraw, or in any way to mitigate the charges in his complaint now exhibited."

I looked over the charges, and made my defence article by article—and when I had gone through the whole, I observed to the committee, that if I had not answered clearly to their satisfaction, if they would put any question that they thought would throw more light on the subject, I would answer it without equivocation or mental reservation. The chairman General Ward made me a low bow, and asked no question, by which they appeared satisfied.

They then took the complaint, and examined the witnesses, one by one upon each charge—"What do you know of this Mr such a one?" (reading the charge) "nothing"—and so to the next, and throughout the whole charges, and the whole witnesses, when "nothing" was the universal answer, except from Marshall Jenkins—He began to tell what the refugees told him at the vineyard when they returned from Nantucket.

The chairman stopped him, and said, "were you at Nantucket?" He answered "no"—"Then you can give no evidence."

One charge against me was corresponding with the enemy. This correspondence was a letter given to Ebenezer Coffin, addressed to General Prescot, requesting the release of his son, and assuring him that he had not been in an armed vessel.

This same Ebenezer acquainted his brother Alexander that I had furnished him with this letter, who used his utmost against me.

The committee were composed of the two branches, the House and Senate, who reported to their several departments. The House then took the vote and cleared us fully except one single vote. But the Senate in their vote held us—The House then desired the Senate to appoint a committee of conference, and they would do the same. They met, and reported, each branch adhering to their former vote.

Thus we were discharged by the House, and held by the Senate, (but not detained) and so it remained until we were set at liberty by the peace.

In the complaint I was charged with being the means of preventing a brig which had much of Jenkins's property on board from being retaken—I suppose that was the case, by reasoning with the owners of the vessel then present.

The seven armed vessels had now gone over the bar, and anchored, waiting for the flowing of the tide to take the brig out—It was then suggested by some hot-headed men, that they could retake her—I admitted it, but asked the owners if it would be an even stake, observing "they have now got what they will take at this time, and if this vessel is stopped, it will bring the whole seven armed vessels into the harbour again, and no doubt the destruction of the town will be the consequence"—For there was no effective force to prevent it. "If you (the owners) will let her go, I am willing to contribute to the loss of vessel and goods on board, in the same proportion that I should pay in a tax of equal amount"—A great number of people were present, who generally united in the proposal. The owners let the vessel go, and I contributed seven hundred and twenty dollars toward the loss of the property, which was more than double my proportion of a like tax.

When this circumstance was known while we were in Boston, it raised great indignation against Jenkins, that such a charge should be in the complaint, when I had made double compensation to what I ought.

In a conversation at the time of our examination with him, several others present, I understood him that I ought to make some concession. My answer was "if turning my hand over by way of concession, would withdraw the complaint, I will never do it—If my innocence will not protect me, and my life should be taken, my blood will be required at thy hands"—This shocked him very much, but it did not last long, as he told some of his friends that he believed Samuel Starbuck and myself were clear. They then asked him why he did not take our names out of the complaint—he replied "because it suits me best to keep them in."

So callous a heart I hope is not often to be met with, thus playing with our lives as with a tennis ball. I am glad to leave this tragic scene and proceed

Sometime in the year 1780 Admiral Arbuthnot returned to England, and Admiral Digby succeeded him. As soon as Arbuthnot was gone, those plundering refugees were upon us again, our protection having ceased by his departure. This renewed our perplexity. The town was convened to consult about measures to prevent destruction The result was to send a committee again to New York, to solicit an order from Admiral Digby similar to that which we had before. It was proposed for me to go with two others. I had then been confined nearly nine months with the rheumatism, had just left my crutches, and was hobbling about with a cane—therefore I could not think of such an

WILLIAM ROTCH

undertaking. But all others utterly refused to go, unless I would accompany them.

This brought a great straight on my mind—to go I thought I could not, and to omit it seemed almost inevitable destruction. At last I consented, under great apprehension that I should not live to return. We accordingly set sail, and when we were off Rhode Island, I was obliged to have them go to the East side of the island, and lay there several days, for my pain was so great that I could not bear the motion of the vessel But we got safe to New York in a few days after it abated.

Admiral Digby had arrived Commodore Affleck (since Admiral) still being there, and he having kindly assisted in getting the permits for a few whaling vessels the year before, we applied first to him. We asked him to introduce us to the Admiral, and assist us in procuring protection against their cruisers in our harbour, and some permits for the fishery.

He looked very stern, and said, "I don't know how you can have the face to ask any indulgence of us—I assisted in getting permits for you last year, which I have been very sorry for. I find that you have abused the confidence we placed in you, for Captain —— who cruised in Boston Bay and its vicinity told me that he could hardly find a vessel but what had the permits, and you deserve no favour" &c &c.—I heard him patiently through, while he was giving us such a lecture, knowing I could overthrow it all I then answered "Commodore Affleck thou hast been greatly imposed upon in this matter. I defy Captain —— to make such a declaration to my face. Those permits were put into *my* hands I delivered them, taking receipts for each, to be returned to me at the end of the voyage, and an obligation that no transfer should be made, nor copies given. I received back all the permits except two before I left home, and should probably have received those two on the day that I sailed. Now if any such duplicity has been practiced, I am the person who is accountable, and I am now here to take the punishment such perfidy deserves."

He immediately became placid, and said, "You deserve favour. I am now going to the Admiral—do you be there in an hour"— We attended punctually He introduced us to the Admiral, and informed him that his predecessor Admiral Arbuthnot granted the people of Nantucket a few permits for the fishery last year, adding, "and I can assure your Excellency they have made no bad use of them."

Thus after a storm came a pleasant calm. We obtained an order, as heretofore, respecting the property in our harbour, and twenty four permits for the fishery—And I returned home much improved in my health.

It was necessary to secrete these documents from American cruisers, but such was the difficulty of distinguishing them, that two were presented to American armed vessels, who immediately took the vessels as prizes. This occasioned us to pursue other means for the security of this small privilege, though a very useful one to us, which I shall mention hereafter.

We were now brought into the most eminent danger, which no human effort could check, much less prevent. Nothing short of the interposition of Divine Providence preserved us from apparent ruin. Several sloops of war, and a number of transports intended paying us a destructive visit. They were in sight of us in the daytime three days, near Cape Poag (Martha's Vineyard)—They got under way three mornings successively, and stood for the island with a fair wind, which each morning soon came round against them, and the tide by that time became unfavourable, which obliged them to return to their anchorage still in view of us.

Before they could make the fourth attempt, orders came for their return to New York for some other expedition.

Thus we were mercifully relieved for that time, after more fearful apprehension than any we had before witnessed. Messengers were arriving one after another, and twice I was called up in the night with the disagreeable information that they were at hand. A solemn time indeed it was, and can never be obliterated from my memory while life and reason are vouchsafed.

We had a few restless spirits amongst us, who were continually involving us in perplexity whenever opportunity offered. From a misrepresentation it was sometimes charged upon the inhabitants at large, though without the least foundation, therefore this Armament was prepared to strip us of what could be found.

When this misrepresentation was discovered, those who authorized the expedition appeared very glad that it was not executed.

I was one with ten men, and two women friends, captured in going to our quarterly meeting at Sandwich, by a British privateer from New York. They had just before taken a cedar boat, and ordered us to depart in it immediately, having first plundered us of what money they could find, but they took neither baggage nor provisions from us.

The vessel was mine, and I pleaded earnestly for her, and sometimes nearly obtained a majority to give her to us—But another can of grog would be stirred up by those who would not consent to release her, and this never failed to gain several to their side. They repeatedly ordered us into the boat and to be gone, but we refused, still pleading for our vessel, 'till the captain of the privateer called to the prize master, to know why he did not send us away. He replied "they will not go."

He then sent a furious fellow to drive us off. Samuel Starbuck and myself were standing together, he approached us with a violent countenance, and uplifted cutlass, saying "Begone into the boat, or I'll cut your heads off." I looked him earnestly in the face, eye to eye, and with a pretty stern accent, said "I am not afraid of thy cutting *my* head off—We are prisoners, treat us as such, and not talk of cutting our heads off."—He dropped his arm with his cutlass, and seemed very much struck at my boldness.

There were now two vessels coming rapidly in pursuit of them, and we thought it was time to be off. They soon retook our vessel, and pursued the privateer, and took her, but the men left her in their boat, and got on shore on the vineyard. They hunted them, and took all except that one who threatened to cut off our heads, and he made his escape.

Our vessel being retaken, I recovered her by paying salvage, as did a young man the most of his money, who had two hundred dollars taken from him.

I now return to the permits granted us by Admiral Digby. The American cruisers generally had knowledge of our whaling vessels having them, therefore every deception and disguise was resorted to, to entrap them. They were too successful in drawing the permit from two and taking them as prizes.

It was now evident that we could proceed no further without having permits from both contending powers. Accordingly the town was convened, and Samuel Starbuck and myself were sent to Congress, to represent our distressed situation, an endeavour to obtain their permission, as well as that of the British for a few vessels.

We set off in mid-winter and arrived in Philadelphia where Congress was sitting. We opened our business first to General Lincoln, Samuel Osgood, Nathaniel Gorham, and Thomas Fitzsimmons. The first was Minister of War, the others were members of Congress. The last a great commercial man. To them we opened our whole business We drew up a Memorial but did not present it until we had an opportunity of stating our case, to the most influential members. Among them was President Madison, who as well as others, treated us with great civility, and seemed to take an interest in our sufferings.

We went to one of the Massachusetts Members, who resided in Boston. He was extremely prejudiced against us. I fell in with him alone, and conversed about two hours with him, endeavouring to impress him with our situation, and the necessity of our having the aid of Congress, but apparently with little effect. At last I asked him three questions, which were "is the whale fishery worth preserving to this Country?"—"Yes"—"Can it be preserved in the present state of things by any place except Nantucket?"—"No"—"Can we pursue it unless you and the British will both give us Permits?"—"No"—"Then pray where is the difficulty?"—Thus we parted. We reported this conversation to our beforementioned friends.

We had now drawn our Memorial, and desired them to look it over. They approved it, and advised us to get the same person to present it. Accordingly we repaired to his apartments, requested him to examine it, and give us his judgment whether our statement appeared correct. He approved it—We then requested him to present it to Congress, if it was agreeable to him to do so—He accepted, and presented it accordingly. It was deliberated upon in Congress, and a disposition appeared to give their aid in its accomplishment. They eventually granted us permits for thirty five vessels for the Whale Fishery.

They were delivered to us, and the next day a vessel arrived from Europe, bringing a rumour of a provincial treaty of peace having been signed by our ministers and the British government, to take place when the peace between England and France should be concluded. And it was not long before an official account of it reached Philadelphia.

Thus ended this destructive war, with the separation of the United States from Great Britain.

Our arduous labours, after five or six weeks attention, were now terminated, and might have been spared, if we had apprehended peace had been so near. The British were still to hold New York, and other territories now ceded to the United States, for a limited time. I obtained liberty to proceed to that city to accomplish some business, and then returned home.

The happy return of peace was now enjoyed in the United States, but poor Nantucket, whose distresses did not end with the war, though rejoiced at the event, still seemed doomed for a time to ruin in the Whale Fishery. Separated from Great Britain, the only market of any consequence for sperm oil, we were necessarily brought under the alien duty of 18 pounds sterling per ton—A duty laid upon aliens to encourage British subjects. Such we then were, but this duty had its full force on us. Sperm oil was sold at Nantucket after the peace at 17 pounds sterling per ton, which before we were separated was worth nearly 30 pounds sterling. 25 pounds sterling was necessary at that time to cover the expenses, and leave a very moderate profit to the owners. Thus a loss of nearly 8 pounds sterling per ton attended the business.

We continued it for two years at a certain loss, with a hope that some more favourable turn might take place. But no such prospect appearing, and the loss I had sustained by captures in the Revolutionary War (about $60,000) had so reduced my property, that I found it necessary to seek some new expedient to prevent the loss of all. I found no probable alternative but to proceed to England, and endeavour to pursue the Fishery from there.

I accordingly took passage in the ship *Maria*, William Mooers Master, accompanied by my son Benjamin, and sailed from Nantucket on the 4th of 7th month 1785. We had a fine passage of twenty three days, five of which, having Easterly winds, we gained only one day's sail forward in that time. I proceeded to London, calling on my old friend Doctor William Cooper at Rochester, (who with his family went to England in this same ship two years before) and requesting him to accompany me to London, which he kindly did.

When we reached Shooter's Hill, in full view of London, and eight miles distant, forcibly feeling the great distance which separated me from my family, myself a stranger in that land, the occasion that drew me there, and the uncertainty of its answering any valuable purpose, I was overwhelmed with sorrow, and my spirits so depressed, that in looking toward that great city, no pleasant pictures were presented to my view. But I found it would not do to give way to despondence, reason resumed her empire, I was there, and something must be attempted.

We reached London, and I took lodgings for myself and my son Benjamin at Thomas Wagstaff's in Gracechurch Street. Our first journey was to the West of England, in which we had the agreeable company of my friend James Phillips. We visited the sea coast from Southampton, to Falmouth, in search of a good situation for the Whale Fishery, if we should conclude to form an establishment on that island. We found several ports suitable for the purpose, but none

The ship *Maria*, in which William Rotch sailed to London in 1785

that we preferred to Falmouth. In that large harbour, there are several smaller, that would do well for the business. I had very favourable offers of divers places, but I was only on discovery, and did not wish to entangle myself with any. After viewing the coast, and spending some days at Plymouth, we took a circuitous route, and returned to London. At Bristol I visited the grave of my brother Joseph, who died there eighteen years before.

My next object was to know what encouragement we could obtain from the British Government.

My friend Robert Barclay perceiving what my business was, spoke to Harry Beaufoy, a Member of Parliament who introduced me to the Chancellor of the Exchequer (the great William Pitt then about twenty seven years of age.)

He received me politely, and heard me patiently. I laid before him our ruinous situation, saying "when the War begun, we declared against taking any part in it, and strenuously adhered to this determination, thus placing ourselves as a neutral island. Nevertheless you have taken from us about two hundred sail of vessels, valued at 200,000 pounds sterling, unjustly and illegally. Had that war been founded on a general declaration against America, we should have been included in it, but it was predicated on a rebellion, consequently none could be in Rebellion but such as were in arms, or those that were aiding such. We have done neither. As a proof of our being without the reach of your declaration, you sent commissioners to restore peace to America, in which any Province, County, town &c. that should make submission, and receive pardon, should be reinstated in their former situation. As we had not offended, we had no submission to make, nor pardon to ask—and it is certainly very hard if we do not stand on better ground than those who have offended, consequently we remained a part of your dominions until separated by the peace."

This last sentence I pressed closely, wherever I could with propriety introduce it, knowing it was a material point.

After I had done he paused some time, and then answered to our remaining a part of their dominions until separated by the peace "most undoubtedly you are right Sir—Now what can be done for you?"

I told him that in the present state of things, the principal part of our inhabitants must leave the island.—Some would go into the country—"A part" said I "wish to continue the Whale Fishery, wherever it can be pursued to advantage—Therefore, my chief business is to lay our distressed situation, and the cause of it, before this Nation, and to ascertain if the fishery is considered an object worth giving such encouragement for a removal to England, as the subject deserves."

Thus our conversation ended, and I withdrew with my friend H. Beaufoy.

The Chancellor of the Exchequer could not be expected to attend to all applications, and I suppose he laid mine before the Privy Council, as the Secretary of the Council Stephen Cotterel sent me a note soon after this conversation, saying the Council would sit at an early day, when they would hear what I had to offer. I waited for that *early* day a month, and then applied to Secretary Cotterel to know what occasioned the delay—The answer was, that so much business lay before them, that they had not been able to attend to it, but would soon.

Thus I waited, not daring to leave town lest I should be called for. This state of things continued more than four months, during which time I received several, what I called unmeaning court messages, "that they were sorry they had not been able to call for me" &c.—

I then desired them to appoint some person for me to confer with, that the matter might be brought to a close. This was done—But unhappily Lord Hawksbury was the person. A greater enemy to America, I believe, could not be found in that body, nor hardly in the Nation.

I waited on him, and informed him what encouragement I thought would induce a removal, which I estimated at 100 pounds-sterling transportation for a family of five persons, and 100 pounds settlement. Say 20,000 pounds—for a hundred families. "Oh!" says he, this is a great sum, and at this time when we are endeavouring to economise in our expenditures."

I replied, "Thou mayst think it a great sum for this Nation to pay, *I* think two thirds of it a great sum for you to have taken from me as an individual, unjustly and illegally." We had a long conversation, and I left him to call again, which I did in a few days.

I then added to my demand the liberty to bring in thirty American ships for the fishery. "Oh no," said he, "that cannot be, our carpenters must be employed," I mentioned that we had some vessels that we possessed before the war—"Those can surely be admitted"

"No—they must be British built."

"Will it be any advantage, if an emigration takes place, for the emigrants to bring property with them?"

"Yes"

"If they can invest their property in articles that will be worth double here to what they are there, will that be an additional advantage to this country?"

"Yes—certainly"

"Then why not bring ships, when two of ours will not cost more than one of yours?"

"Oh we don't make mercantile calculations, 't is seamen we want."

"Then surely two of our ships will answer your purpose better than one of yours, as they will make double the number of seamen, which must be the very thing aimed at." He saw that he was in a dilemma, which he could not reason himself out of, and struggled through with some violence.

He had now made his nice calculation of 87 pounds—10 for transportation, and settlement of a family—and says he, "I am about a fishery bill, and I want to come to something that I may insert it" &c.

My answer was, "Thy offer is no object, therefore go on with thy fishery bill, without any regard to me."

I was then taking leave, and withdrawing

"Well, Mr. Rotch, you'll call on me again in two or three days."

"I see no necessity for it."

"But I desire you would"

"If it is thy desire perhaps I may call."

However, he let me rest but one day before he sent for me. He had the old story over again, but I told him it was unnecessary to enter again into the subject. I then informed him that I had heard a rumour that Nantucket had agreed to furnish France with a quantity of oil. He stepped to his bureau, took out one of a file of papers, and pretended to read an entire contradiction, though I was satisfied there was not a line there on the subject.

I said, "it was only a vague report that I heard, and I cannot vouch for the truth of it—but we are like drowning men, catching at every straw that passes by, therefore I am now determined to go to France, and see what it is—if there is any such contract, sufficient to retain us at Nantucket, neither you, nor any other nation shall have us, and if it is insufficient, I will endeavour to enlarge it."

"Ah!" says he, "Quakers go to France?"

"Yes," I replied, "but with regret." I then parted with Lord Hawksbury for the last time.

I immediately embarked with my son for Dunkirk, where I drew up our proposals, and sent them to Paris, not wishing to proceed further, until I found the disposition of the French court. They sent for us to come immediately—we lost no time in answering the summons, and soon reached Paris. The Master of Requests who was the proper Minister to receive our proposals, and to make his remarks on the several articles, had examined them, and made his remarks accordingly. The propositions were

1st A full and free enjoyment of our religion, according to the principles of the people called Quakers—

To which he annexed,

"*Accordé.*"

2nd An entire exemption from military requisitions of every kind.

To this he annexed the following just remark, "as they are a peaceable people, and meddle not with the quarrels of princes, neither internal nor external, this proposition may be granted."

The other propositions related to the regulation of the Whale Fishery.

We next proceeded to the several other ministers at Versailles, five in number. First to Calone, Comptroller of Finance. We gave our reasons for not taking off our hats on introduction, to them all. Calone replied, "I care nothing about your hats, if your hearts are right." Next, to the aged Vergennes, Minister of Foreign affair. Then to the Marshall DeCastre, Minister of Marine. To the Prince of Reubec *Generalissimo* of Flanders—and last to the *Intendant* of Flanders, who all agreed to my proposals.

We then returned to Paris, and were to visit Versailles again, to take leave, according to the etiquette of the court. Before we set off one of the ministers asked us if we did not wish to see the palace. We excused ourselves, as we did not think curiosity would justify us, if our plain way would give any offence. While we remained in Paris we received a note, saying the minister had spoken to the king who gave full liberty for the Nantucket Friends (avoiding the name of Quakers when they found that it was given us in reproach) to visit the palace, both its public and private apartments, when he was out (which was almost every day).

To view the private apartments was a great indulgence, not often granted except to persons of note. But unfavourably for us, the king did not happen to be out on the day we went to take leave, which was a disappointment, but we went through the public apartments, and into the chapel. When we hesitated at the latter, the officer insisted on our entering in our own way, showing us everything remarkable, and pointing out the places occupied by the Royal family in time of mass, &c.

We now took leave and returned to London.

After I was gone to France Lord Hawksbury became alarmed, and enquired of Harry Beaufoy if I was gone to France—he answered in the affirmative .

"Why is he gone there?"

"For what you or any other person would have gone—You would not make him an offer worth his acceptance He will now try what can be done in France."

Alexander Champion wrote to me, I suppose at Lord Hawksbury's request, to inform me that he had made provision in his Fishery Bill for us, and inserted liberty to bring in forty ships, instead of thirty which I demanded, he having forgotten the number, but it was too late.

This letter was brought to our apartments, and we understood the bearer to enquire, if a Dutch gentleman resided there—he was answered in the negative, and my letter was lodged in a small letter office, always an appendage to the large hotels. The very evening we left Paris it was brought to me.

We now returned to London, and I was soon sent for by George Rose (I suppose father of the minister lately sent to the United States) who was one of Pitt's secretaries. He enquired if I had contracted with France. I told him, no—I did not come to make any contract. Propositions were the extent of my business "You are then at liberty to agree with us—and I am authorised by Mr Pitt to tell you that you shall make your own terms."

I told him it was too late—"I made very moderate proposals to you, but could not obtain anything worth my notice I went to France, sent forward my proposals, which were doubly advantageous to what I had offered your Government They considered them but a short time, and on my arrival in Paris were ready to act. I had a separate interview with all the Ministers of State necessary to the subject (five in number) who all agreed to, and granted my demands. This was effected in five hours, when I had waited to be called by your Privy Council more than four months."

He still insisted that as I was not bound to France, I should make my own terms with *them*, but all in vain—the time had passed over.

The Chase—Sperm-whaling scene

Lord Sheffield also sent for me on the same subject, but was soon convinced that it was too late. The minority came to me for materials to attack Lord Hawksbury, but I refused.

I now began to prepare for returning to my family. Accordingly I bought a good ship, and with William Mooers Master, we left the Downs the 11th of 10 month 1786—After a tremendous passage of sixty eight days, in which we had twelve heavy storms, we arrived in Boston, and by way of Providence and Newport reached my own home on the 1st of 1st month 1787, and to my unspeakable comfort found my family well after an absence of eighteen months.

We next prepared to increase our fishery in Dunkirk, and my son Benjamin returned there, to superintend the business, he having become a partner with my son-in-law Samuel Rodman & myself.

After remaining at home nearly four years, I thought it best to make another voyage, to assist my son in our business at Dunkirk—And not expecting to return in less than three years, a term too long to be separated from my family, I proposed to my wife to go with me, and take our daughters Lydia & Mary with us, to which she consented—and we also took with us my son Benjamin's wife and child.

We sailed from New Bedford in the ship *Maria & Eliza*, which I bought for the purpose, Abisha Haydon Master, on the 29th of 7th month 1790. We arrived at Dunkirk in thirty eight days, and found our son Benjamin in health, and greatly rejoiced to receive his wife and child, as well as to see us.

Early in the year 1791 I was called upon with my son to attend the National Assembly at Paris. We were joined by John Marsillac in presenting a petition to that body for some privileges and exemptions connected with our religious principles.

The petition was drawn by John Marsillac before we reached Paris, and notice given that it must be presented the next day.

On perusing it, we found some material alterations necessary. And in some instances it was difficult to express in French the alterations we made in English, without losing their force. My not understanding the French language it was impossible to have such expressions inserted as I thought necessary

And the time was so short, that we were obliged to let it pass with much fewer amendments than we wished.

The hour was come for presenting it, and the previous notice given of the Quaker petition, I suppose drew every member in town to his seat. The galleries for spectators were filled, and many could not be accommodated, nor did we wonder at their curiosity, considering the novelty of the object.

We had been, with Brissot De Warville, Clavier, and some others looking over the petition until the latest moment, and must now proceed to the assembly. They with several others had come to accompany us, and just as we were moving. One observed, "You have no cockades—You must put them on." We told them we could not—It was a distinguishing badge that we could not make use of.

"But," said they, "it is required by Law, to prevent distinction, that people may not be abused, for their lives are in danger without them, and there is always a large body of the lower classes about the assembly that we have to pass through."

Our answer was, that we could not do it, whatever might be the consequence—That we were willing to go as far as we could, and if stopped, we must submit to it—We saw that our friends were full of fear for our safety. We set out under no small apprehension, but trusting to that power that can turn the hearts of men as a water course is turned, we passed through this great concourse without interruption, and reached the waiting room of the assembly.

A messenger informed the president of our arrival, and we were immediately called to the bar.

John Marsillac read the petition, with Brissot at his elbow, to correct him in his emphasis, which he frequently did, unperceived I believe except by us. At the close of every subject, there was a general clapping of hands, and the officers whose business it was, endeavouring to hush them that the reading might proceed, this hushing I thought was hissing, from my ignorance of the language, and apprehended all was going wrong, until better informed.

After the reading was concluded, President Mirabeau read his answer. The clapping was repeated at the end of every subject—at the close, the president said—"The assembly invites you to stay its sitting."

As we were passing to the seats assigned us, a person touched Benjamin, and said, "I rejoice to see something of your principles brought before this assembly." He did not know who it was. After we were seated, several members came to us for conversation on the subject of our principles. We remained until the assembly rose, and then retired to our lodgings.

We next found that a visit to the influential members, in their private hotels, was necessary, to impress them with the reasonableness of our requests. We accordingly proceeded, John Marsillac, Benjamin & myself, and met with a polite reception from all except two, and nothing more than a careless indifference from them. One was Bernard, a young man of good talents, but great vanity. At our approach he offered us no seats, but threw himself on his sofa at great ease, which we were told was his common attitude, when applicants of much more consequence than we were came into his presence. The other was Tallyrand—after endeavouring to impress him with the foundation of our petition, he made no reply, but let us pass silently away.

We generally found a number of persons with the members we visited, not of the assembly, but applicants, soliciting their interest for the different objects they were pursuing—and the features of our petition always led to an opportunity of opening our principles at large, particularly that respecting war. They invariably enquired and listened with great attention, and seldom was any opposition expressed. We had much conversation with Bishop Gregory, who was a very catholic man, liberal in his sentiments, and much esteemed and also with Rabant De St Etienne then Bishop of Autun. He was a very valuable man, and I believe was a blessing to many over whom he presided. He was one that fell a victim to the guillotine under the sanguinary reign of Robespierre—He inclined to converse much on non-resistance, and finally, thus summed up what he considered the view of its advocates, and of pure Christianity—

"If an assassin comes, to take my life, and I conscientiously refrain from taking his to save it, I may trust some interposition for my deliverance. If however, no interposition appearing, I still refrain from precipitating a soul unprepared into eternity, and he is suffered to effect his purpose on me, I may hope to find mercy for myself."

The object of our petition was of little consequence to me, whether granted or not compared with the opportunity we now had, of somewhat spreading the knowledge of our fundamental principles, above all that of the Inward Light or Spirit of God in every man, as a primary rule of faith and practice. We met with a number of serious persons, who were in great measure convinced of the rectitude of our faith, and they gathered to us at our hotel one evening after another, one inviting others to come with them, until these social meetings in our apartments became exceedingly interesting. The conversation was almost wholly on religious subjects, and they always appeared well satisfied with the hours thus spent.

It was then a turbulent time in Paris, and much more so after-

CAPTURING A SPERM WHALE

wards—Several of those valuable persons fell in the Reign of Terror, and others are beyond my knowledge, but the remembrance of those evenings and of the feeling of divine influence that attended them I believe will never pass away.

One of our visitors informed us that the Duchess of Bourbon was greatly interested in the principles we profess, and said if we wished to see her, he would make way for it—but Benjamin's business calling him home, and my not speaking the French language, we could not accept the proffered interview. We therefore returned to Dunkirk.

In the course of the year 1792 fresh trials awaited us. A great insurrection took place in Dunkirk, founded on a rumour of the exportation of corn—Several houses were attacked, their furniture totally destroyed, and the families, among whom were particular friends of ours, but just escaped with their lives. At last the military were called in aid of the civil authority, and fifteen of the rioters were killed before they were quelled. The head of one of these families escaped in disguise, and his wife and daughters were secretly conveyed to our house at midnight, whence before daybreak a friend escorted them on their journey to the *chateau* of her father sixty miles distant. Martial law was proclaimed, and wherever five men were seen together in the evening and night orders were given to fire upon them. It was indeed an awful time.

A great trial now assailed us, which I had anticipated with serious apprehension.—that of an illumination for the victories of the French over the Austrians. The illumination was announced as for tomorrow evening. Having very little time to consider what could be done, Benjamin and myself thought best to go immediately to the mayor and magistrates then assembled, to inform them that we could not illuminate, and the cause. That as we could take no part in war, we could not join in rejoicings for victory. On opening the subject, they were much alarmed for our safety, and asked us what protection they could afford us. We replied "that is no part of our business—we only wish to place our refusal on the right ground, and to remove any apprehension that we are opposing the government."

"Well," said the mayor, "keep to your principles—your houses are your own—the streets are ours—and we shall pursue such measures as we think proper for the peace of this town." We retired, though not without some fear that they would send an armed force. Should this be the case, and any life be lost in endeavouring to protect us, I thought it would be insupportable. However they took another method, and sent men to erect a frame before our house, and three other houses occupied by those of our denomination, and hang a dozen lamps upon it.

The mayor had also the great kindness to have a similar frame with lamps, placed before his own house, in addition to the usual full illumination; and he once, and the magistrates several times walked by our house, to see if all remained quiet—for they were under great apprehension. The evening being fine, and great numbers walking in the streets, they generally stopped to enquire why this singular kind of illumination, when they were informed by the person placed there by the mayor for this purpose, and to take charge of the lights. On his assuring them that we were not opposed to the government, but were Quakers, they went on their way. We had all withdrawn into a back parlour where we spent the evening, and thus passed this trying occasion unmolested.

A circumstance took place in the afternoon previous to the illumination, which I believe contributed in part to our remaining quiet. My son was passing in the street, and observed a number of men conversing very earnestly. One said, "If there are any aristocrats who do not illuminate they will be destroyed."

Benjamin then remarked to him, that he hoped that would not be the criterion to judge aristocrats by, as he could not illuminate, and gave his reasons. The man who had been so earnest then addressed him thus—"I am glad I know your reasons, and I will endeavour all in my power to prevent your being injured."

Another of the company said, "Mr Rotch, this man can do more with those people whom you have the most reason to fear, than any man in this town"—and I have no doubt that he used his influence with those very people. Thus we may frequently see a concurrence of circumstances in our preservation, which is by many attributed to chance I believe it is rather the watchful care of our Heavenly Father, however undeserving we may be.

Another illumination took place soon after, when the same course was pursued towards us by the mayor as before. A young man was passing our house late in the evening, when many lights in the town were extinguished, and saw two men searching on the ground. On enquiring what they were seeking, they said, "We are looking for something to demolish these windows—they are aristocrats, and do not illuminate." He told them they must not molest us, that we were no aristocrats, but were Quakers, whose religious principles forbade public rejoicings on any occasion, and persuaded the men away—of this the young man's father informed us the next morning.

The next illumination was on the arrival of commissioners sent from Paris to stir up the people to action. My son being absent, I requested Louis DeBacque to go with me to the commissioners, and as my interpreter, give our reasons for taking no part in it.

We found them in one of the forts, and after Louis had communicated what I wished, the principal among them came to me, and taking me by the hand, desired we would do nothing contrary to our scruples on their account.

After some further friendly expressions, he turned to a large body of people present, and thus addressed them—"We are now about establishing a government on the same principles that William Penn the Quaker established Pennsylvania—and I find there are a few Quakers in this town, whose religious principles do not admit of any public rejoicings, and I desire they may not be molested."

That same evening the commissioners assembled the town at the town house, to address them on their public affairs, and in the course of it, he took up our case again, and as before, desired we might not be molested, but protected. We afterwards found that several candles had been stuck around some pillars on the outside of our house, though we did not know it at the time.

This must have been done by mechanics in our employ, several of whom on each of these occasions, passed much of the evening in walking before our houses to see if there was any service they could render, and telling those who enquired that we were Quakers and not aristocrats.

In the beginning of 1793 I became fully aware that war between England and France would soon take place. Therefore it was time for me to leave the country, in order to save our vessels if captured by the English. I proceeded to England, two of them were captured, full of oil, and condemned, but we recovered both by my being in England, where I arrived two weeks before the war took place.

My going to France to pursue the Whale Fishery so disappointed Lord Hawksbury, that he undertook to be revenged on me for his own folly, and I have no doubt gave directions to the cruisers to take any of our vessels that they met with going to France. When the *Ospray* was taken by a King's ship, the officer who was sent on board to examine her papers, called to the captain, and said, "You'll take this vessel in sir, she belongs to Mr Rotch."

My wife and family embarked secretly from Dunkirk with many Americans in a ship bound to America, and were landed in England, where I had the great comfort of receiving them four months after I left them in France.

Louis Sixteenth was guillotined two days after I left that agitated country—an event solemnly anticipated, and deeply deplored by many who dared not manifest what they felt.

We were now settled in London, where we enjoyed the company of many old acquaintances and friends until the summer of 1794. My son William sent the ship *Barclay*, David Swain, Master, to France with a cargo, and ordered her from thence to London to take us to America.

We embarked the 24th of 7th month, had a long passage of sixty one days, and arrived in Boston 23rd of 9th month 1794. The night before our arrival an awful circumstance took place during a squall—Calvin Swain, brother of the captain, fell from the main top sail yard into the long boat, and was instantly killed.

We soon proceeded to New Bedford, and after spending a few days there, returned to our home at Nantucket, finding all our children, and grandchildren well that we left more than four years before, and six added in Samuel's and William's families.

We staid a year at our old habitation, and then removed to New Bedford, where we have remained until now, when I have entered on my eightieth year.

Many occurrences I omit in giving the foregoing account, or they would swell this scrip to a considerable volume—When I take a retrospective view of this part of my life, of the dangers to which I have been exposed, and the numerous preservations I have witnessed, to be attributed to nothing but that Superintending Power, who is ever ready to succour the workmanship of his holy hand, it fills me with astonishment and admiration, and seeing my own unworthiness, I may exclaim with the psalmist, "*What is man that thou art mindful of him, or the son of man that thou visitest him!*"

New Bedford 2nd mo. 1814.

NEW BEDFORD ABOUT 1800
SHOWING WILLIAM ROTCH'S MANSION, AND WILLIAM ROTCH HIMSELF IN HIS CHAISE, THE ONLY PRIVATE CARRIAGE THEN KEPT IN THE TOWN.

The Respectful Petition
OF THE
CHRISTIAN SOCIETY OF FRIENDS
CALLED QUAKERS,
DELIVERED BEFORE THE NATIONAL ASSEMBLY,
Thursday 10th February 1791.

Respectable Legislators:

The French nation having appointed you her legislators, and your hearts having been disposed to enact wise laws, we solicit the extension of your justice and benevolence to the society of peaceable Christians to which we belong.

You know that in several states of Europe and North America, there are a great number of Christians known by the name of Quakers, who profess to serve God according to the ancient simplicity of the primitive Christian church. Several towns and villages of Languedoc contain a number of families attached to this primitive Christianity. Many other families, which came from America, have settled at Dunkirk, under the auspices of the late government, in consequence of the invitation given to the inhabitants of Nantucket, for the purpose of extending the French fisheries. These islanders have proved themselves worthy of your kindness by their success, and the same motive will induce them to continue to deserve it. Concerns, however, of far greater moment, have this day brought us before you.

In an age signal for the increase of knowledge, you have been struck with this truth, that conscience, the immediate relation of man with his creator, cannot be subject to the power of man: and this principle of justice hath induced you to decree a general liberty for all forms of worship. This is one of the noblest decrees of the French legislature.

You have set a great example to the nations which continue to persecute for religion, and sooner or later, we hope, they will follow it.

We are come to implore this spirit of Justice, that we may be suffered, without molestation, to conform to some principles, and to use some forms, to which the great family of Friends called Quakers, have been inviolably attached ever since their rise.

Great persecutions have been inflicted on us, on account of one of these principles, but to no purpose. Providence hath enabled us to surmount them, without using violence. We mean the principle which forbids us to take arms, and kill men on any pretence; a principle consistent with the holy scriptures: "*render not*" (said Christ) "*evil for evil, but do good to your enemies.*"

Would to Heaven this principle were universally adopted! All mankind becoming one family, would be brethren united by acts of kindness. Generous Frenchmen, you are convinced of its truth; you have already begun to reduce it to practice; you have decreed never to defile your hands with blood in pursuit of conquest. This measure brings you, it brings the whole world, a step towards universal peace. You cannot therefore behold with an unfriendly eye men who accelerate it by their example. They have proved in Pennsylvania, that vast establishments may be formed, raised, and supported without military preparations, and without shedding human blood.

We submit to your laws, and only desire the privilege of being here, as in other countries, the brethren of all men—never to take up arms against any. England and the United States of America, where our brethren are far more numerous than in France, allow us peaceably to follow this great principle of our religion, nor do they esteem us useless members of the community. We have another request to make, which we hope you will not refuse us; because it flows from those principles of justice to which you do homage. In our registers of births, marriages and burials, we have preserved the simplicity of the primitive Church. Our maxims forbid useless forms, and limit us to those which are necessary for ascertaining the terms of human life, consistently with the good order of society. We request that our simple registers may be deemed sufficient to legalize our marriages and births, and authenticate our deaths, by causing a declaration thereof to be made before a magistrate—

Finally, we request that we may be exempted from all oaths, Christ having expressly forbidden them in these words, "*You have heard that it hath been said by them of old time, thou shalt perform thine Oaths; but I say unto you swear not at all, but let your yea be yea, and your nay nay."*

Wise legislators, you are persuaded as well as we, that an oath is no assurance of sincerity; that it can give no additional force to the declaration of an honest man, and does not deter a perjurer. You admit that an oath is but a peculiar way of making a declaration,—as it were a peculiar mode of speech. We hope therefore you will not refuse to hear us in ours. It is that of our common Master—that of Christ. We trust that we shall not be suspected of a wish to evade the great purpose of the civic oath. We are earnest to declare in this place, that we will continue true to the constitution which you have formed; we cherish and respect it, and it is our full purpose to follow its laws in all their purity; on the other hand, if our words, if our evidences are found to be false, we willingly submit to the penalties on false witnesses and perjurers.

Can you, respectable legislators, hesitate to grant our request? Cast your eyes on the history of our Society, in the countries in which we are established. More than a century hath elapsed, and we have never been found in any conspiracy against the government. Our temperate rule of life forbids ambition and luxury, and the purpose of our watchful discipline is to preserve us in the practice of those manners, to which we were led by the exhortations and example of our founder.

We esteem employment a duty enjoined on all: and this persuasion renders us active and industrious. In this respect therefore our Society may prove useful to France. By favouring us you encourage industry. Industry now seeks those countries where the honest industrious man will be under no apprehensions of seeing the produce of a century of labour snatched away, in an instant, by the hand of persecution.

Now that France is becoming the asylum of liberty, of equal law and of brotherly kindness, and adds to these sources of prosperity, perfect liberty for every individual to obey the dictates of his conscience, in relation to the Almighty;—what prospects of advantage will arise to induce our brethren who inhabit less happy climes, to settle in France, a country favoured by nature, as soon as they learn that you have granted them the same civil and religious liberty which they enjoy in England and the United States of America.

Such is the respectful petition we present to you, for the relief of our brethren in France, and for the good of a country which we love. We hope among your important engagements in reforming this great empire, and multiplying the sources of its happiness, you will extend your justice and regard to us and our children: it will bring upon you the reward of the Almighty, and the love of virtuous men.

Answer of the President

Quakers, who have fled from persecutors and tyrants, cannot but address with confidence those legislators who have, for the first time in France, made the rights of mankind the basis of law. And France, now reformed, France in the bosom of peace, (which she will always consider herself bound to revere, and which she wishes to all other nations) may become another happy Pennsylvania.

As a system of philanthropy, we admire your principles. They remind us that the origin of every society was a family united by its manners, its affections, and its wants, and doubtless those would certainly be the most sublime institutions, which would renew the human race, and bring them back to this primitive and virtuous original.

The examination of your principles, as a matter of opinion, no longer concerns us: we have decided on that point. There is a kind of property which no man would put into the common stock: the motions of his soul, the freedom of his thought. In this sacred domain, man is placed in a hierarchy far above the social state. As citizen, he must adopt a form of government:—but as a thinking being, the universe is his country.

As principles of religion, your doctrines will not be the subject of our deliberation. The relation of every man with the Supreme Being is independent of all political institutions. Between God and the heart of man what government would dare to interpose?—

As civil maxims, your claims must be submitted to the discussion of the legislative body. We will examine whether the forms you observe in order to ascertain births and marriages, be sufficient to authenticate those descents which the division of property renders indispensable, independently of good customs.

We will consider whether a declaration, subject to the penalties against false witnesses and perjury, be not in fact an oath.

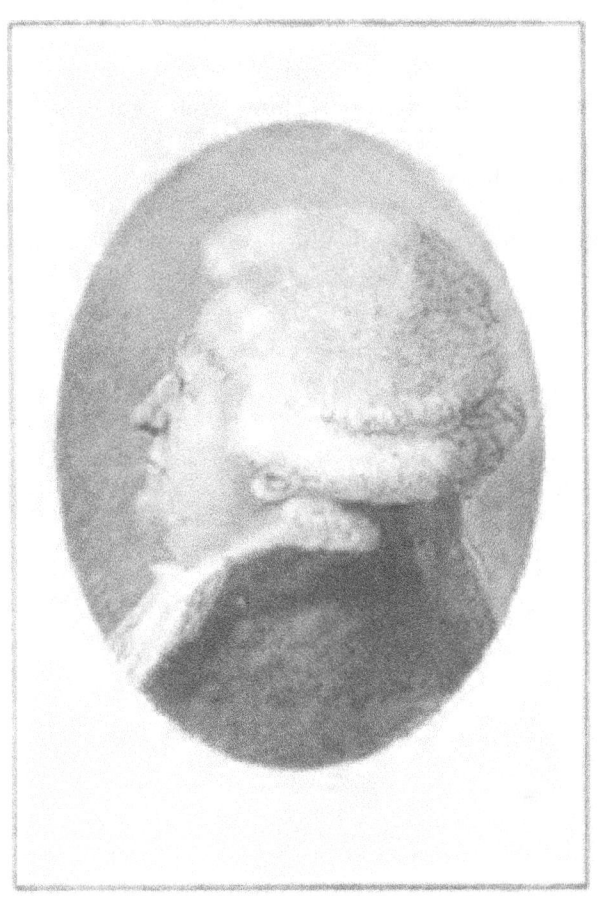

Mirabeau

Worthy citizens, you have already taken that civic oath which every man deserving of freedom hath thought a privilege rather than a duty. You have not taken God to witness, but you have appealed to your consciences. And is not a pure conscience a Heaven without a cloud? Is not that part of man a ray of the divinity? —

You also say that one of your religious tenets forbids you to take up arms, or to kill, on any pretence whatsoever. It is certainly a noble philosophical principle, which thus does a kind of homage to humanity. But consider well, whether the defence of yourselves, and your equals, be not also a religious duty? You would otherwise be overpowered by tyrants!—Since we have procured liberty for you, and for ourselves, why should you refuse to preserve it?

Had your brethren in Pennsylvania been less remote from the savages, would they have suffered their wives, their children, their parents to be massacred rather than resist? And are not stupid tyrants, and ferocious conquerors also savages?

The assembly will, in its wisdom, consider all your requests. But whenever *I* meet a Quaker, I shall say.

My brother, if thou hast a right to be free, thou hast a right to prevent anyone from making thee a slave.

As thou lovest thy fellow-creature, suffer not a tyrant to destroy him: it would be killing him thyself.

Thou desirest peace—but consider—weakness invites war—general resistance would prove an universal peace.

The assembly invites you to stay its sitting.

Copy of Thomas Jenkins's Complaint

To the Honourable the Council, and the Honourable the House of Representatives, in General Court assembled, at Boston, November 1779.

Thomas Jenkins humbly sheweth—That as a true and liege subject of the State of the Massachusetts Bay, as well as from enormous personal injuries received, he is most strongly urged to lay the following representation and complaint before the supreme legislature of the State.

Your petitioner complains of Dr Benjamin Tupper, Timothy Folger Esqr, William Rotch, Samuel Starbuck, and Kezia Coffin, all of the island of Nantucket, as persons dangerous, and inimical to the freedom and independence of this and the other United States of America; as encouragers, aiders and abettors of the enemy, in making inroads on the state territories, and depredations on the property of the good subjects of this State.

It can be clearly proved (if your honours should think fit to order an enquiry) that reiterated attempts have been made by some of the above persons, to induce the inhabitants of said island, to settle correspondence with, and openly join the enemy.

In particular the said Timothy Folger applied to the selectmen of the town of Sherbourn, in a written request, subscribed with his own hand, to call a convention of the town, in order to choose a committee to treat with the British commanders at New York and Rhode Island; and also whether it was expedient for the town any longer to pay taxes to this State; and upon the said application being reprobated by the select men as highly prejudicial and inimical to the honour and interest of the State, said Folger declared that several of the principal inhabitants of the town were of his opinion; and then insolently told the select men they deserved to be damned if they refused to comply with his proposal.

This treasonable proposition will be proved by the select men. That there was a private correspondence carried on by some or all of the aforesaid persons with the enemy can be proved by the annexed list of witnesses No 2—and that the commander of the predatory fleet which came to Nantucket last spring, confessed that they never should have come there on the design they executed, had they not been repeatedly called upon and invited for the purpose, by the friends to the British Government, in the infamous number of whom the persons above complained of were notoriously enrolled. Doctor Samuel Gelston will prove this confession.

That upon the arrival of the Renegado Fleet at the bar of the harbour, the aforesaid Rotch and Folger together with one Josiah Barker, (without any appointment or consent of the town) went on board the said fleet, and after tarrying sometime, came on shore in company with several of the principal refugee officers, and immediately went to the said Rotch's house, where, after calling in three or four other men of the same inimical character with themselves, a long consultation was held.

In about an hour the council broke up, and one of the said officers with a number of his people proceeded immediately to some warehouses of your complainant, and robbed him of—260 barrels of sperm oil, 1800 lbs of whale bone, 2300 wt. of iron, 1200 lbs of coffee, 20,000 wt of tobacco, and a number of smaller articles, all which they carried off, together with a brig, one moiety of which he owned, to the loss of your complainant, twenty five hundred guineas at least.— This property was pointed out to them by the said Dr Tupper and Kezia Coffin. That other effects of the true and liege subjects of this State were particularly pointed out to the enemy for plunder, by some of the persons above complained of, and especially by said Starbuck, can be amply proved.

Your complainant begs leave further to add, that after the enemy had got possession of his brig above mentioned, frightened with a false alarm, they precipitately left the harbour, and the said brig behind them, with only five men in her; upon which some of the well disposed inhabitants proposed securing her, which might with ease have been effected; but the said Folger and Rotch with some others of the same complexion and sentiments interposed and dissuaded, and opposed the intention of the people; by means of which the said brig and cargo were finally carried off, after a pilot was procured by the enemy, who was induced to take charge of the vessel by the advice of the aforesaid William Rotch. To put the inimical and treasonable sentiments and designs of the said Dr Tupper beyond all dispute, after he had returned from New York, with said Starbuck and Rotch, where they had gone on an illegal and dangerous errand, upon a town meeting being convened, said Tupper after having menaced and abused all those persons, who had been opposed to the said triumvirate going to New York, moved that a committee should be chosen, for the purpose of seeing that the king's servants, meaning the adherents and officers of the British king, should receive suitable respect and protection and be kindly used, and that all refractory persons, (meaning such liege subjects of this State as were opposed to their traitorous proceedings) should be apprehended and sent where they would meet their punishment.

Innumerable other instances of the most dangerous and illegal conduct in some or all the persons now complained of can be produced, should your honours think fit to order an enquiry to be made; which your petitioner and complainant humbly prays may be ordered, as well for the public interest, as that some reparation may be made him, and his other suffering brethren, who have sustained very heavy losses, by the cruel and treasonable management of those people; and that such order may issue from your honours as shall compel the persons charged as above, to answer to these articles of complaint, and that summonses may be granted for the witnesses whose names are herewith handed to your honours, to attend at such time as your honours shall order the enquiry to be made.

And your petitioner as in duty bound shall ever pray.——

    (Signed)   Thomas Jenkins.

a true copy
Attest John Avery DC.J.——

Witnesses to the several charges *viz.*

Timothy Fitch	Marshall Jenkins
(*Medford*)	(*M Vineyd*)
Benjn Folger	Paul Pinkham
Walter Folger	Benjn Hussey
Shubael Barnard	Stephen Hussey

Peter Macy
Ebenezer Coffin
Dr. Samuel Gelston
John Waterman
Shubael Downs
 (*Walpole*)
Francis Chase

Seth Jenkins
Shubael Worth
Stephen Fish
William Hammett
John Ramsole
George Hussey 2nd

Copy

ALSO FROM LEONAUR
AVAILABLE IN SOFTCOVER OR HARDCOVER WITH DUST JACKET

A HISTORY OF THE FRENCH & INDIAN WAR *by Arthur G. Bradley*—The Seven Years War as it was fought in the New World has always fascinated students of military history—here is the story of that confrontation.

WASHINGTON'S EARLY CAMPAIGNS *by James Hadden*—The French Post Expedition, Great Meadows and Braddock's Defeat—including Braddock's Orderly Books.

BOUQUET & THE OHIO INDIAN WAR *by Cyrus Cort & William Smith*—Two Accounts of the Campaigns of 1763-1764: Bouquet's Campaigns by Cyrus Cort & The History of Bouquet's Expeditions by William Smith.

NARRATIVES OF THE FRENCH & INDIAN WAR: 2 *by David Holden, Samuel Jenks, Lemuel Lyon, Mary Cochrane Rogers & Henry T. Blake*—Contains The Diary of Sergeant David Holden, Captain Samuel Jenks' Journal, The Journal of Lemuel Lyon, Journal of a French Officer at the Siege of Quebec, A Battle Fought on Snowshoes & The Battle of Lake George.

NARRATIVES OF THE FRENCH & INDIAN WAR *by Brown, Eastburn, Hawks & Putnam*—Ranger Brown's Narrative, The Adventures of Robert Eastburn, The Journal of Rufus Putnam—Provincial Infantry & Orderly Book and Journal of Major John Hawks on the Ticonderoga-Crown Point Campaign.

THE 7TH (QUEEN'S OWN) HUSSARS: Volume 1—1688-1792 *by C. R. B. Barrett*—As Dragoons During the Flanders Campaign, War of the Austrian Succession and the Seven Years War.

INDIA'S FREE LANCES *by H. G. Keene*—European Mercenary Commanders in Hindustan 1770-1820.

THE BENGAL EUROPEAN REGIMENT *by P. R. Innes*—An Elite Regiment of the Honourable East India Company 1756-1858.

MUSKET & TOMAHAWK *by Francis Parkman*—A Military History of the French & Indian War, 1753-1760.

THE BLACK WATCH AT TICONDEROGA *by Frederick B. Richards*—Campaigns in the French & Indian War.

QUEEN'S RANGERS *by Frederick B. Richards*—John Simcoe and his Rangers During the Revolutionary War for America.

AVAILABLE ONLINE AT **www.leonaur.com**
AND FROM ALL GOOD BOOK STORES

ALSO FROM LEONAUR
AVAILABLE IN SOFTCOVER OR HARDCOVER WITH DUST JACKET

JOURNALS OF ROBERT ROGERS OF THE RANGERS *by Robert Rogers*—The exploits of Rogers & the Rangers in his own words during 1755-1761 in the French & Indian War.

GALLOPING GUNS *by James Young*—The Experiences of an Officer of the Bengal Horse Artillery During the Second Maratha War 1804-1805.

GORDON *by Demetrius Charles Boulger*—The Career of Gordon of Khartoum.

THE BATTLE OF NEW ORLEANS *by Zachary F. Smith*—The final major engagement of the War of 1812.

THE TWO WARS OF MRS DUBERLY *by Frances Isabella Duberly*—An Intrepid Victorian Lady's Experience of the Crimea and Indian Mutiny.

WITH THE GUARDS' BRIGADE DURING THE BOER WAR *by Edward P. Lowry*—On Campaign from Bloemfontein to Koomati Poort and Back.

THE REBELLIOUS DUCHESS *by Paul F. S. Dermoncourt*—The Adventures of the Duchess of Berri and Her Attempt to Overthrow French Monarchy.

MEN OF THE MUTINY *by John Tulloch Nash & Henry Metcalfe*—Two Accounts of the Great Indian Mutiny of 1857: Fighting with the Bengal Yeomanry Cavalry & Private Metcalfe at Lucknow.

CAMPAIGN IN THE CRIMEA *by George Shuldham Peard*—The Recollections of an Officer of the 20th Regiment of Foot.

WITHIN SEBASTOPOL *by K. Hodasevich*—A Narrative of the Campaign in the Crimea, and of the Events of the Siege.

WITH THE CAVALRY TO AFGHANISTAN *by William Taylor*—The Experiences of a Trooper of H. M. 4th Light Dragoons During the First Afghan War.

THE CAWNPORE MAN *by Mowbray Thompson*—A First Hand Account of the Siege and Massacre During the Indian Mutiny By One of Four Survivors.

BRIGADE COMMANDER: AFGHANISTAN *by Henry Brooke*—The Journal of the Commander of the 2nd Infantry Brigade, Kandahar Field Force During the Second Afghan War.

BANCROFT OF THE BENGAL HORSE ARTILLERY *by N. W. Bancroft*—An Account of the First Sikh War 1845-1846.

AVAILABLE ONLINE AT www.leonaur.com
AND FROM ALL GOOD BOOK STORES

ALSO FROM LEONAUR
AVAILABLE IN SOFTCOVER OR HARDCOVER WITH DUST JACKET

AFGHANISTAN: THE BELEAGUERED BRIGADE *by G. R. Gleig*—An Account of Sale's Brigade During the First Afghan War.

IN THE RANKS OF THE C. I. V *by Erskine Childers*—With the City Imperial Volunteer Battery (Honourable Artillery Company) in the Second Boer War.

THE BENGAL NATIVE ARMY *by F. G. Cardew*—An Invaluable Reference Resource.

THE 7TH (QUEEN'S OWN) HUSSARS: Volume 4—1688-1914 *by C. R. B. Barrett*—Uniforms, Equipment, Weapons, Traditions, the Services of Notable Officers and Men & the Appendices to All Volumes—Volume 4: 1688-1914.

THE SWORD OF THE CROWN *by Eric W. Sheppard*—A History of the British Army to 1914.

THE 7TH (QUEEN'S OWN) HUSSARS: Volume 3—1818-1914 *by C. R. B. Barrett*—On Campaign During the Canadian Rebellion, the Indian Mutiny, the Sudan, Matabeleland, Mashonaland and the Boer War Volume 3: 1818-1914.

THE KHARTOUM CAMPAIGN *by Bennet Burleigh*—A Special Correspondent's View of the Reconquest of the Sudan by British and Egyptian Forces under Kitchener—1898.

EL PUCHERO *by Richard McSherry*—The Letters of a Surgeon of Volunteers During Scott's Campaign of the American-Mexican War 1847-1848.

RIFLEMAN SAHIB *by E. Maude*—The Recollections of an Officer of the Bombay Rifles During the Southern Mahratta Campaign, Second Sikh War, Persian Campaign and Indian Mutiny.

THE KING'S HUSSAR *by Edwin Mole*—The Recollections of a 14th (King's) Hussar During the Victorian Era.

JOHN COMPANY'S CAVALRYMAN *by William Johnson*—The Experiences of a British Soldier in the Crimea, the Persian Campaign and the Indian Mutiny.

COLENSO & DURNFORD'S ZULU WAR *by Frances E. Colenso & Edward Durnford*—The first and possibly the most important history of the Zulu War.

U. S. DRAGOON *by Samuel E. Chamberlain*—Experiences in the Mexican War 1846-48 and on the South Western Frontier.

AVAILABLE ONLINE AT www.leonaur.com
AND FROM ALL GOOD BOOK STORES

LEONAUR
ALSO FROM LEONAUR
AVAILABLE IN SOFTCOVER OR HARDCOVER WITH DUST JACKET

THE 2ND MAORI WAR: 1860-1861 by Robert Carey—The Second Maori War, or First Taranaki War, one more bloody instalment of the conflicts between European settlers and the indigenous Maori people.

A JOURNAL OF THE SECOND SIKH WAR by Daniel A. Sandford—The Experiences of an Ensign of the 2nd Bengal European Regiment During the Campaign in the Punjab, India, 1848-49.

THE LIGHT INFANTRY OFFICER by John H. Cooke—The Experiences of an Officer of the 43rd Light Infantry in America During the War of 1812.

BUSHVELDT CARBINEERS by George Witton—The War Against the Boers in South Africa and the 'Breaker' Morant Incident.

LAKE'S CAMPAIGNS IN INDIA by Hugh Pearse—The Second Anglo Maratha War, 1803-1807.

BRITAIN IN AFGHANISTAN 1: THE FIRST AFGHAN WAR 1839-42 by Archibald Forbes—From invasion to destruction-a British military disaster.

BRITAIN IN AFGHANISTAN 2: THE SECOND AFGHAN WAR 1878-80 by Archibald Forbes—This is the history of the Second Afghan War-another episode of British military history typified by savagery, massacre, siege and battles.

UP AMONG THE PANDIES by Vivian Dering Majendie—Experiences of a British Officer on Campaign During the Indian Mutiny, 1857-1858.

MUTINY: 1857 by James Humphries—Authentic Voices from the Indian Mutiny-First Hand Accounts of Battles, Sieges and Personal Hardships.

BLOW THE BUGLE, DRAW THE SWORD by W. H. G. Kingston—The Wars, Campaigns, Regiments and Soldiers of the British & Indian Armies During the Victorian Era, 1839-1898.

WAR BEYOND THE DRAGON PAGODA by Major J. J. Snodgrass—A Personal Narrative of the First Anglo-Burmese War 1824 - 1826.

THE HERO OF ALIWAL by James Humphries—The Campaigns of Sir Harry Smith in India, 1843-1846, During the Gwalior War & the First Sikh War.

ALL FOR A SHILLING A DAY by Donald F. Featherstone—The story of H.M. 16th, the Queen's Lancers During the first Sikh War 1845-1846.

AVAILABLE ONLINE AT **www.leonaur.com**
AND FROM ALL GOOD BOOK STORES

ALSO FROM LEONAUR
AVAILABLE IN SOFTCOVER OR HARDCOVER WITH DUST JACKET

THE FALL OF THE MOGHUL EMPIRE OF HINDUSTAN by *H. G. Keene*—By the beginning of the nineteenth century, as British and Indian armies under Lake and Wellesley dominated the scene, a little over half a century of conflict brought the Moghul Empire to its knees.

LADY SALE'S AFGHANISTAN by *Florentia Sale*—An Indomitable Victorian Lady's Account of the Retreat from Kabul During the First Afghan War.

THE CAMPAIGN OF MAGENTA AND SOLFERINO 1859 by *Harold Carmichael Wylly*—The Decisive Conflict for the Unification of Italy.

FRENCH'S CAVALRY CAMPAIGN by *J. G. Maydon*—A Special Correspondent's View of British Army Mounted Troops During the Boer War.

CAVALRY AT WATERLOO by *Sir Evelyn Wood*—British Mounted Troops During the Campaign of 1815.

THE SUBALTERN by *George Robert Gleig*—The Experiences of an Officer of the 85th Light Infantry During the Peninsular War.

NAPOLEON AT BAY, 1814 by *F. Loraine Petre*—The Campaigns to the Fall of the First Empire.

NAPOLEON AND THE CAMPAIGN OF 1806 by *Colonel Vachée*—The Napoleonic Method of Organisation and Command to the Battles of Jena & Auerstädt.

THE COMPLETE ADVENTURES IN THE CONNAUGHT RANGERS by *William Grattan*—The 88th Regiment during the Napoleonic Wars by a Serving Officer.

BUGLER AND OFFICER OF THE RIFLES by *William Green & Harry Smith*—With the 95th (Rifles) during the Peninsular & Waterloo Campaigns of the Napoleonic Wars.

NAPOLEONIC WAR STORIES by *Sir Arthur Quiller-Couch*—Tales of soldiers, spies, battles & sieges from the Peninsular & Waterloo campaingns.

CAPTAIN OF THE 95TH (RIFLES) by *Jonathan Leach*—An officer of Wellington's sharpshooters during the Peninsular, South of France and Waterloo campaigns of the Napoleonic wars.

RIFLEMAN COSTELLO by *Edward Costello*—The adventures of a soldier of the 95th (Rifles) in the Peninsular & Waterloo Campaigns of the Napoleonic wars.

AVAILABLE ONLINE AT **www.leonaur.com**
AND FROM ALL GOOD BOOK STORES

ALSO FROM LEONAUR
AVAILABLE IN SOFTCOVER OR HARDCOVER WITH DUST JACKET

AT THEM WITH THE BAYONET *by Donald F. Featherstone*—The first Anglo-Sikh War 1845-1846.

STEPHEN CRANE'S BATTLES *by Stephen Crane*—Nine Decisive Battles Recounted by the Author of 'The Red Badge of Courage'.

THE GURKHA WAR *by H. T. Prinsep*—The Anglo-Nepalese Conflict in North East India 1814-1816.

FIRE & BLOOD *by G. R. Gleig*—The burning of Washington & the battle of New Orleans, 1814, through the eyes of a young British soldier.

SOUND ADVANCE! *by Joseph Anderson*—Experiences of an officer of HM 50th regiment in Australia, Burma & the Gwalior war.

THE CAMPAIGN OF THE INDUS *by Thomas Holdsworth*—Experiences of a British Officer of the 2nd (Queen's Royal) Regiment in the Campaign to Place Shah Shuja on the Throne of Afghanistan 1838 - 1840.

WITH THE MADRAS EUROPEAN REGIMENT IN BURMA *by John Butler*—The Experiences of an Officer of the Honourable East India Company's Army During the First Anglo-Burmese War 1824 - 1826.

IN ZULULAND WITH THE BRITISH ARMY *by Charles L. Norris-Newman*—The Anglo-Zulu war of 1879 through the first-hand experiences of a special correspondent.

BESIEGED IN LUCKNOW *by Martin Richard Gubbins*—The first Anglo-Sikh War 1845-1846.

A TIGER ON HORSEBACK *by L. March Phillips*—The Experiences of a Trooper & Officer of Rimington's Guides - The Tigers - during the Anglo-Boer war 1899 - 1902.

SEPOYS, SIEGE & STORM *by Charles John Griffiths*—The Experiences of a young officer of H.M.'s 61st Regiment at Ferozepore, Delhi ridge and at the fall of Delhi during the Indian mutiny 1857.

CAMPAIGNING IN ZULULAND *by W. E. Montague*—Experiences on campaign during the Zulu war of 1879 with the 94th Regiment.

THE STORY OF THE GUIDES *by G.J. Younghusband*—The Exploits of the Soldiers of the famous Indian Army Regiment from the northwest frontier 1847 - 1900.

AVAILABLE ONLINE AT **www.leonaur.com**
AND FROM ALL GOOD BOOK STORES

www.ingramcontent.com/pod-product-compliance
Lightning Source LLC
Chambersburg PA
CBHW021008090426
42738CB00007B/709